Synthetic Organic Chemistry and the Nobel Prize Volume 1

The Nobel Prize is science's highest award, as is the case with non-science fields too, and it is therefore arguably the most internationally recognized award in the world. This unique set of volumes focuses on summarizing the Nobel Prize within organic chemistry, as well as the specializations within this specialty. Any reader researching the history of the field of organic chemistry will be interested in this work. Furthermore, it serves as an outstanding resource for providing a better understanding of the circumstances that led to these amazing discoveries and what has happened as a result in the years since.

Series: Synthetic Organic Chemistry and the Nobel Prize

Synthetic Organic Chemistry and the Nobel Prize Volume 1

Author – Dr. John D'Angelo

Synthetic Organic Chemistry and the Nobel Prize Volume 1

John G. D'Angelo

CRC Press
Taylor & Francis Group
Boca Raton London New York

CRC Press is an imprint of the
Taylor & Francis Group, an **informa** business

First edition published 2023
by CRC Press
6000 Broken Sound Parkway NW, Suite 300, Boca Raton, FL 33487-2742

and by CRC Press
4 Park Square, Milton Park, Abingdon, Oxon, OX14 4RN

© 2023 Taylor & Francis Group, LLC
CRC Press is an imprint of Taylor & Francis Group, LLC

ISBN: 9780367438975 (hbk)
ISBN: 9781032417202 (pbk)
ISBN: 9781003006831 (ebk)

DOI: 10.1201/9781003006831

Typeset in Times
by Deanta Global Publishing Services, Chennai, India

Contents

Author   vii

Introduction   1
History   1
Other Facts   2
  How Is the Nobel Prize Selected?   9
  Why Laureates?   12
  The Medal, Diploma, and Cash Prize   12
  Nobels Proven Wrong?   13
  Nazis and the Disappearing Nobel Prizes   14
  Controversies and Snubs   15
    Controversies   15
  Nobel Snubs   16
    Women and the Nobel   18
  Race and the Nobel Prize   20
  What to do?   29
  The Chemistry Prize   32
    Future of the Nobel Prize   33
    Future of the Chemistry Nobel Prize   34
  Predictions   34
  The Prize and Society   38
About This Series of Books   38
  Honorable mentions   40
General sources   41
Cited references   42

1 Fischer   43
What Was the Discovery and Who Made the Discovery?   43
Summary   49
General sources   49
Cited references   49

2 Wallach   51
Summary   54
General sources   54

3 Sabatier and Grignard   **55**
Grignard   60
Summary   67
General sources   67
Cited references   67

4 Diels and Alder   **69**
Summary   86
General sources   86
Cited references   86

5 Woodward   **89**
Controversies   93
Quinine Synthesis   93
Hoffmann and Corey   94
Select Syntheses   94
Summary   97
General sources   98
Cited references   98

Index   99

Author

photo credit: Kerry Kautzman
D'Angelo, 2022)

Dr. D'Angelo, earned his BS in Chemistry from the State University of New York at Stony Brook in 2000. While at Stony Brook, he worked in Prof. Peter Tonge's lab on research toward elucidating the mechanism of action of FAS-II inhibitors for anti-*mycobacterium tuberculosis* drugs. While there, he was an active member of the chemistry club, serving as its treasurer for a year. After graduating, he worked as a summer research associate at Stony Brook in Prof. Nancy Goroff's lab, working toward the synthesis of molecular belts. He then earned his PhD from the University of Connecticut in 2005, working in the laboratories of Michael B. Smith. There, Dr D'Angelo worked on the synthesis of 2-nucleobase, 5-hydroxymethyl lactams as putative anti-HIV agents while also investigating the usefulness of the conducting polymer poly-(3,4-ethylinedioxy thiophene) as a chemical reagent. He served as a teaching assistant during most of his four years at UCONN and was awarded the Outstanding TA award during one of these years. After completing his PhD, he took a position as a postdoctoral research associate at The Johns Hopkins University in Prof. Gary H. Posner's lab. There, Dr. D'Angelo worked on the development of artemisinin derivatives as anti-malarial and anti-*toxoplasma gondii* derivatives. In 2007, Dr. D'Angelo accepted a position at Alfred University at the rank of Assistant Professor and in 2013, he was awarded tenure and promotion to the rank of Associate Professor at Alfred and awarded promotion to Professor in July 2021. Dr. D'Angelo's research has continued to focus on the chemical reactivity of conducting polymers and has been expanded to pedagogical research and scientific ethics. He served as the local ACS section (Corning) chair in 2014 and in 2021, and as the Faculty Senate president for two consecutive terms, serving in this capacity from 2014 to 2018 and became Chair of the Chemistry Division at Alfred in 2021. He is also the author of four books. One, on scientific misconduct, is in its second edition, and a second book on scientific misconduct is intended to be a workbook with hypothetical cases that students can work through. A third book, written with his PhD advisor, outlines a process for using the chemical search engine Reaxsys to teach reactions, and the fourth book is an organic chemistry textbook published through the web-based publisher Top Hat. He is also an author of 13 peer-reviewed publications (three in his independent career) and two patents. This four-volume series on organic chemistry and the Nobel Prize is his latest authoring endeavor.

Introduction

HISTORY

The Nobel Prize—a prize recognized at least in name—as one of the, if not THE, premier rewards for genius is arguably the most famous award in the world. It is unlikely that someone past a High School education has never heard of the Nobel Prize. Awarded (mostly) annually since 1901 in the subjects of Chemistry, Physics, Physiology or Medicine, Peace, and Literature and joined in 1968 by the Sveriges Riksbank Prize in Economic Sciences in memory of Alfred Nobel (Figure 0.1), these prizes carry a medal, diploma, and cash prize for those chosen for this high honor in addition to the accompanying recognition. It is important to note that the Economic Sciences prize is, formally speaking, not a Nobel Prize, though they're awarded at the same time and this prize is treated very much like a Nobel.

The prize was created by Alfred Nobel in 1895 in his last will and testament, with the largest share of his considerable fortune allocated to the series of prizes. The "rules" set out in Nobel's will about the award are still at least mostly adhered to today, over a century later. The chief difference is that although Nobel stipulated that the award should be made for a scientific matter from the *preceding year*, it appears (to me anyway) that it has more recently become more of a lifetime achievement award of sorts, at least in the life and physical sciences awards (Chemistry, Physiology or Medicine, and Physics). Exactly when this started is difficult to pin down, but it is very clearly the current modus operandi.

The primary source of Nobel's wealth was as the inventor of dynamite, a stabilized form of the explosive nitroglycerine. In a very real way, the establishment of the prize was its own first controversy. Initially, the creation of the prize caused somewhat of a scandal, and it wasn't until several years (1901) after his death (1896) that his requests were finally fulfilled with the first series of awards. His own family opposed the prize; much of his considerable wealth had been bequeathed for its creation rather than to them, so it is easy to understand their objection. In his will, he specifically called for the Swedish Academy of Sciences to award both the Physics and Chemistry awards; the Karolinksa Institute in Stockholm to award the Physiology or Medicine award;

DOI: 10.1201/9781003006831-1

FIGURE 0.1:   Artist rendition of Alfred Nobel.

the Peace prize to be awarded by a committee of five to be selected by the Norwegian Storting; and the literature award to be awarded by the Academy in Stockholm.

One may wonder why Alfred Nobel created the prize at all. Although there is no direct evidence to support the claims, legend has it that Alfred was horrified by an errant obituary—his own—mistakenly published upon the death of his *brother*. In it, Alfred was referred to as "the merchant of death," due to how his invention, dynamite, had been used. In addition to less violent uses (e.g. mining), dynamite and other modern derivatives/analogs are also used as a weapon. Being labeled a "merchant of death" is enough to rattle just about anyone's emotional cage. The timing (1895) of his will (his third one) seems to fit this legend since the version of his last will and testament establishing the award was signed seven years after the death of his brother, rather than before his brother's death while his wealth had come far earlier.

OTHER FACTS

The awards are traditionally announced over a one-week period in early October. The ceremony and lectures are held in Stockholm later in the year, usually in December. As of this writing, in mid-2022, since its inception, the Nobel Prize has been awarded to 947 individual laureates and 28 organizations. Since great intellect and achievements often are not isolated, four individuals have won more than one Nobel Prize and three organizations have done likewise. Furthermore, some Nobel greatness apparently "runs" in some families. Among these, the Curie family is by far the most successful though they are not alone in having more than one family member earn an award.

Although there are no rules limiting how many awards one recipient can win, no award can be given for more than two works (i.e., topics) in any given year, and the award may only be shared, whether it is for one work or two, between three recipients among each of the awarded fields. In some cases, this sharing will happen by more than one topic earning the honor, other times, it will be due to more than one individual or organization contributing to the same work. In short, the following permutations are the only ones allowed.

- One topic, one person
- One topic, two people
- One topic, three people
- Topic A one person and Topic B one person
- Topic A one person and Topic B two people

Prior to 1974, posthumous awarding of the prize was permitted and happened twice. Dag Hammarskjöl (1961, Peace) and Erik Axel Karlfeldt (1931, Literature) were both posthumously awarded the Nobel Prize. Both died earlier in the year they were awarded their prize and likely were at the very least nominated prior to their deaths. Hammarskjöl, the second Secretary-General of the United Nations was given the award "for developing the UN into an effective and constructive international organization, capable of giving life to the principles and aims expressed in the UN Charter." Hammarskjöl died in a plane crash in September 1961. Karlfeldt, meanwhile, was given the award for "the poetry of Erik Axel Karlfeldt" and died in April 1931. One exception since 1974 has been made for extenuating circumstances. The committee did not know that the awardee, Ralph Steinman (2011, Physiology or Medicine) had passed away a mere three days before announcing the award. Steinman was given the award "for his discovery of the dendritic cell and its role in adaptive immunity."

The maximum number of recipients stipulated has become a modern source of controversy. Science, and in fact the world, is far more collaborative than it was in Nobel's time. Rarely—effectively never—does a modern researcher make a major discovery alone. Instead, most modern work involves literal scores of individuals playing an important even if small role in creating the final mosaic. Thus, restricting the Nobel Prize to such a small number guarantees that people contributing to the work are left out. It is even safe to say it leaves out *most* of the people contributing. That it is inevitably and arguably often the people doing the lion's share of actual work only magnifies the problem. This is a fair criticism of the prize. However, we must recognize that it is also fair to lay this claim against *any* award. Even if it is more noticeable because of the magnitude of *this* prize, all awards (including sports awards) that recognize an *individual* ignore the contributions of essential supporting role players. By no means am I trying to justify or defend this reality. I only wish to point out that all awards can be so criticized. This is covered in more detail, as are some potential resolutions, in the controversies section later in this introduction.

Nobel's will also stipulates what should happen in years where no award is given for a field. The will states that the prize money is to be reserved for the following year, and if even then no award can be made, the funds are added to the foundation's restricted funds. This (no award for a field) has happened, between the awards, a total of 49 times, mostly during times of war. Technically, the statutes for skipping a year refer to it being possible to not make an award if "none of the works under consideration is found to be of the importance indicated in the first paragraph" but that the award was consistently not awarded during times of global conflict is probably not a coincidence. It is unlikely that somehow, no important work was done during those years. Literature was skipped in 2018 amid controversy involving one of the committee members though in this case, it was awarded the next year, making delayed more accurate in this case than skipped. Peace is the one "skipped" the most at 19 times. The others do not even measure to half, in order: Medicine—9; Chemistry—8; Literature—7; and Physics—6. The award for Economic Sciences has not existed during any times of global conflict and has never been skipped.

(Perhaps) surprisingly, two Nobel laureates declined the prize: Jean-Paul Sartre (1964, Literature) on the grounds that he consistently declined *all* official honors and Le Duc Tho (1973, Peace) along with U.S. Secretary of State (who accepted the award), Henry Kissinger. Although they were jointly awarded the prize for their work on the Vietnam Peace accord, Tho pointed to the ongoing situation in Vietnam as justification for declining. Four others were *forced* to decline the ward. Three of the four were Germans—Richard Kuhn (1938, Chemistry) "for his work on carotenoids and vitamins," Adolf Butenandt (1939, Chemistry) "for his work on sex hormones," and Gerhard Domagk (1939, Physiology or Medicine) "for the discovery of the antibacterial effects of prontosil"—forbidden from accepting the award by Hitler; all four later were able to receive the diploma and medal but not the prize. The fourth, Boris Pasternak (1958, Literature) "for his important achievement both in contemporary lyrical poetry and the field of the great Russian epic tradition," a Russian, initially accepted his award but was later coerced by authorities to decline it.

There are also no restrictions regarding the awardee being a free person; three laureates were imprisoned at the time of the award. Carl von Ossietzky (1935, Peace), Aung San Suu Kyi (1991, Peace), and Liu Xiaobo (2010, Peace) were all awarded the Prize while incarcerated. von Ossietzky was given the award "for his burning love for freedom of thought and expression and his valuable contribution to the cause of peace" and was an anti-Nazi who revealed the rearmament efforts of Germany in violation of the Versailles Treaty that ended World War I. He was sent to a concentration camp when the Nazis seized power. Hitler's fury in response to von Ossietzky's award led him to prohibit all Germans from receiving the Nobel Prize. Kyi was awarded the prize "for her non-violent struggle for democracy and human rights." She opposed the military junta that ruled Burma, efforts that landed her under house arrest for nearly 15 years. After being released, she resumed her political career only to be arrested again after a military coup and later being sentenced to a total of eight years. Finally, Liu Xiaobo, given the award "for his long and non-violent struggle for fundamental human rights in China," received his sentence for the crime of speaking. His first stint in prison was due to his part in the student protests on Tiananmen Square in 1989 and a second (this time in a labor camp) for his criticism of China's one-party system. Most recently, in 2008, Liu co-authored Charta 08, which advocates for China's shift in the direction of democracy. His charge was undermining the state authorities, and this earned him an 11-year sentence.

Some Nobel laureates were downright deplorable. Take, for example, Dr. D. Carleton Gajdusek (1976, Physiology or Medicine), a pediatrician who discovered the role of prions in a disease known as Kuru, which is related to mad cow disease. He was also a self-admitted and, I dare say, unapologetic pedophile. Many of his victims were also his research patients. Another, Fritz Haber (1918, Chemistry) was potentially a war criminal. Both are covered in more detail in the controversies section of this introduction.

Sometimes, Nobel greatness runs in the family. A list of Nobel Prize-winning families is found in Table 0.1. The Curie family is the most prolific of the "Nobel Families." Pierre Curie won the prize (1903, Physics), sharing it with his wife, Marie (a.k.a. Madame) Curie, who went on to win one of her own (1911, Chemistry) several years later. One of their daughters, Irène Joliot-Curie, and her husband Frédéric Joliot also went on to share a Nobel Prize (1935, Chemistry). This brings their family total to five shared or individual Nobel Prizes. As if this were not enough, although not an actual awardee, Henry R. Labouisse, husband of another of Marie and Pierre's daughters Ève, accepted the prize on behalf of UNICEF (1965, Peace). All told, this one family had a hand in no less than *six* Nobel Prizes. It is extremely unlikely that something like this will *ever* be matched. It appears that the Curies are the New York Yankees of science. The Curies, though the most prolific, do not

TABLE 0.1   List of Nobel Prize-Winning Families

FAMILY	NAME(S)	YEAR	FIELD	TOPIC
Curie	Piere and Marie Curie	1903	Physics	"In recognition of the extraordinary services they have rendered by their joint researches on the radiation phenomena discovered by Professor Henri Becquerel"
	Marie Curie	1911	Chemistry	"in recognition of her services to the advancement of chemistry by the discovery of the elements radium and polonium, by the isolation of radium and the study of the nature and compounds of this remarkable element"
	Irène Joliot-Curie and Frédéric Joliot	1935	Chemistry	"in recognition of their synthesis of new radioactive elements"
	Henry R. Labouisse (on behalf of UNICEF)	1965	Peace	"for its effort to enhance solidarity between nations and reduce the difference between rich and poor states"
Cori	Carl and Gerty Cori	1947	Physiology or Medicine	"for their discovery of the course of the catalytic conversion of glycogen"
Duflo & Banergee	Esther Duflo and Abhijit Banerjee	2019	Economic Sciences	"for their experimental approach to alleviating global poverty"
Moser	May-Britt and Edvard I. Moser	2014	Physiology or Medicine	"for their discoveries of cells that constitute a positioning system in the brain"

(Continued)

TABLE 0.1 (CONTINUED) List of Nobel Prize-Winning Families

FAMILY	NAME(S)	YEAR	FIELD	TOPIC
Mydral	Gunner Mydral	1974	Economic Sciences	"for their pioneering work in the theory of money and economic fluctuations and for their penetrating analysis of the interdependence of economic, social and institutional phenomena"
	Alva Mydral	1982	Peace	"for their work for disarmament and nuclear and weapon-free zones"
Bragg	Sir William and William Lawrence Bragg	1915	Physics	"for their services in the analysis of crystal structure by means of X-rays"
Bohr	Niels	1922	Physics	"for his services in the investigation of the structure of atoms and of the radiation emanating from them"
	Agae	1975	Physics	"for the discovery of the connection between collective motion and particle motion in atomic nuclei and the development of the theory of the structure of the atomic nucleus based on this connection"
Euler-Chelpin	Hons von Euler-Chelpin	1929	Chemistry	"for their investigations on the fermentation of sugar and fermentative enzymes"
	Ulf von Euler	1970	Physiology or Medicine	"for their discoveries concerning the humoral transmittors in the nerve terminals and the mechanism for their storage, release and inactivation"

(Continued)

TABLE 0.1 (CONTINUED) List of Nobel Prize-Winning Families

FAMILY	NAME(S)	YEAR	FIELD	TOPIC
Kornberg	Arthur Kornberg	1959	Physiology or Medicine	"for their discovery of the mechanisms in the biological synthesis of ribonucleic acid and deoxyribonucleic acid"
	Roger Kornberg	2006	Chemistry	"for his studies of the molecular basis of eukaryotic transcription"
Siegbahn	Manne Siegbahn	1924	Physics	"for his discoveries and research in the field of X-ray spectroscopy"
	Kai Siegbahn	1981	Physics	"for his contribution to the development of high-resolution electron spectroscopy"
Thomson	J.J. Thomson	1906	Physics	"in recognition of the great merits of his theoretical and experimental investigations on the conduction of electricity by gases"
	George Thomson	1937	Physics	"for their experimental discovery of the diffraction of electrons by crystals"
Tinbergen	Jan Tinbergen	1969	Economic Sciences	"for having developed and applied dynamic models for the analysis of economic processes"
	Nikolaas Tinbergen	1973	Physiology or Medicine	"for their discoveries concerning organization and elicitation of individual and social behaviour patterns"

hold an exclusive claim to multiple family members earning a Nobel Prize. Other spousal pairs to share a prize are: Carl and Gerty Cori (1947, Physiology or Medicine); Esther Duflo and Abhijit Banerjee (2019, Sveriges Riksbank Prize in Economic Sciences in Memory of Alfred Nobel); and May-Britt and Edvard I. Moser (2014, Physiology or Medicine) all shared a Nobel Prize while Gunnar Mydral (1974, Sveriges Riksbank Prize in Economic Sciences in Memory of Alfred Nobel) and his spouse Alva Mydral (1982, Peace) brought two separate Nobels to their house. Father and son pairs have also brought home Nobel Prizes with one pair: Sir William Henry and son William Lawrence Bragg (1915, Physics) sharing a prize. Other father–son pairs include Niels (1922, Physics) and Agae Bohr (1975, Physics); Hons von Euler-Chelpin (1929, Chemistry) and Ulf von Euler (1970, Physiology or Medicine); Arthur (1959, Physiology or Medicine) and Roger Kornberg (2006, Chemistry); Manne (1924, Physics) and Kai Siegbahn (1981, Physics); and J.J. (1906, Physics) and George Thomson (1937, Physics). Rounding out the keeping it in the family trend are Jan (1969 Sveriges Riksbank Prize in Economic Sciences in Memory of Alfred Nobel) and younger brother Nikolaas Tinbergen (1973, Physiology or Medicine).

How Is the Nobel Prize Selected?

Each prize has a selection committee that, around September of the preceding year, sends confidential forms to individuals considered qualified and competent to nominate. Committee members are all members of the academy and serve a period of three years. Not just anyone can serve as an expert adviser, as well; only those specially appointed.

Across the awards, nominations are not allowed to be revealed until 50 years after the prize has been awarded. However, nominators are under no obligation to keep their nominations confidential. The qualified nominators and timeline (which does have some slight variations) for each award are summarized below.

Chemistry[1]

QUALIFIED NOMINATORS	TIMELINE
Member of Royal Swedish Academy of Sciences. Member of the Nobel Committee for Chemistry or Physics. Nobel laureate in Chemistry or Physics. Permanent professor in the sciences of chemistry at the universities and institutes of technology of Sweden, Denmark, Finland, Iceland, and Norway and Karolinksa Instituet, Stockholm. Holders of corresponding Chairs in at least six universities or university colleges selected by the Academy of Sciences with a view to ensuring appropriate distribution of the different countries and their centers of learning. Other scientists from whom the academy may see fit to invite proposals.	~September previous year: Nomination invitations sent out. January 31: Deadline for nominations. March–June: Consultation with experts. June–August: Report writing with recommendations. September: Academy receives final report. October: Majority vote and announcement. December: Ceremony.

Physics[2]

QUALIFIED NOMINATORS	TIMELINE
Swedish and foreign members of the Royal Swedish Academy of Sciences.	~September previous year: Nomination invitations sent out.
Members of the Nobel Committee for Physics.	
Nobel laureates in Physics.	January 31: Deadline for nominations.
Tenured professors in the Physical sciences at the universities and institutes of technology of Sweden, Denmark, Finland, Iceland and Norway, and Karolinska Instituet, Stockholm.	March–June: Consultation with experts.
Holders of corresponding chairs in at least six universities or university colleges (normally, hundreds of universities) selected by the Academy of Sciences with a view to ensuring the appropriate distribution over the different countries and their seats of learning.	June–August: Report writing with recommendations. September: Academy receives final report. October: Majority vote and announcement.
Other scientists from whom the Academy may see fit to invite proposals.	December: Ceremony.

Physiology or Medicine[3]

QUALIFIED NOMINATORS	TIMELINE
Members of the Nobel Assembly at Karolinska Instituet, Stockholm.	~September previous year: Nomination invitations sent out.
Swedish and foreign members of the Medicine and Biology classes of the Royal Swedish Academy of Sciences.	January 31: Deadline for nominations.
Nobel laureates in Physiology or Medicine and Chemistry.	March–June: Consultation with experts.
Members of the Nobel Committee not qualified under paragraph 1 above.	
Holders of established posts as full professors at the faculties of medicine in Sweden and holders of similar posts at the faculties of medicine or similar institutions in Denmark, Finland, Iceland, and Norway.	June–August: Report writing with recommendations. September: Academy receives final report.
Holders of similar posts at no fewer than six other faculties of medicine at universities around the world, selected by the Nobel Assembly, with a view to ensuring the appropriate distribution of the task among various countries.	October: Majority vote and announcement. December: Ceremony.
Scientists whom the Nobel Assembly may otherwise see fit to approach.	
No self-nominations are considered.	

Literature[4]

QUALIFIED NOMINATORS	TIMELINE
Members of the Swedish Academy and of other academies, institutions, and societies which are similar to it in construction and purpose. Professors of literature and of linguistics at universities and university colleges. Previous Nobel laureates in Literature. Presidents of those societies of authors that are representative of the literary production in their respective countries.	~September previous year: nominations sent out. January 31: Deadline for nominations. April: 15–20 preliminary candidates. May: five final candidates. June–August: reading of productions. September: Academy members confer. October: Award announced. December: Ceremony.

Peace[5]

QUALIFIED NOMINATORS	TIMELINE
Members of national assemblies and national governments (cabinet members/ministers) of sovereign states as well as current heads of states. Members of The International Court of Justice in The Hague and The Permanent Court of Arbitration in The Hague Members of l'Institut de Droit International Members of the international board of the Women's International League for Peace and Freedom. University professors, professors emeriti and associate professors of history, social sciences, law, philosophy, theology, and religion; university rectors and university directors (or their equivalents); directors of peace research institutes and foreign policy institutes. Persons who have been awarded the Nobel Peace Prize. Members of the main board of directors or its equivalent of organizations that have been awarded the Nobel Peace Prize. Current and former members of the Norwegian Nobel Committee (proposals by current members of the Committee to be submitted no later than at the first meeting of the Committee after 1 February) Former advisers to the Norwegian Nobel Committee.	~September previous year: Nomination invitations sent out. January 31: Deadline for nominations. February–April: Preparation of short list. April–August: Adviser review. October: Majority vote and announcement. December: Ceremony.

Economic Sciences[6]

QUALIFIED NOMINATORS	TIMELINE
Swedish and foreign members of the Royal Swedish Academy of Sciences.	~September previous year: Nomination invitations sent out.
Members of the Prize Committee for the Sveriges Riksbank Prize in Economic Sciences in Memory of Alfred Nobel.	January 31: Deadline for nominations.
Persons who have been awarded the Sveriges Riksbank Prize in Economic Sciences in Memory of Alfred Nobel.	March–June: Consultation with experts.
Permanent professors in relevant subjects at the universities and colleges in Sweden, Denmark, Finland, Iceland, and Norway;.	June–August: Report and Recommendation writing.
Holders of corresponding chairs in at least six universities or colleges, selected for the relevant year by the Academy of Sciences with a view to ensuring the appropriate distribution between different countries and their seats of learning; and	September: Academy receives report on finalists.
	October: Majority vote and announcement.
Other scientists from whom the Academy may see fit to invite proposals.	December: Ceremony.

The last point effectively serves as an "out" of sorts to allow for solicitation from anyone in the field. Importantly, nobody can self-nominate. One would imagine (or hope) such hubris is unlikely to be received well by the committee, anyway. Likewise, at least for the chemistry prize it is not possible for *just anyone* to submit a nomination; for example, you cannot nominate one of your pals.

Why Laureates?

The reference to Nobel Prize winners as laureates has its roots in ancient Greece. A laurel wreath (Figure 0.2), a circular crown made of branches and leaves of the bay laurel, was awarded to victors of athletic competitions and poetic meets in ancient Greece as a sign of honor. The term "laureate" to describe awardees of the Nobel Prize is used to harken back to this honor.

The Medal, Diploma, and Cash Prize

Created by Swedish and Norwegian artists and calligraphers, each diploma is quite literally a work of art. The medals (Figure 0.3) all have Nobel's image on them with his birth and death years, though the Economic Sciences prize is

FIGURE 0.2: Laurel Wreath.

FIGURE 0.3: Nobel Medal.

slightly different in its design. Recall that this prize is in Nobel's memory and was not one of the original subjects endowed by Nobel. Each medal is handmade out of 18-carat recycled gold. The cash award is currently set at nine million Swedish Krona. As of this writing, this is equal to ~851,000 Euro and ~965,000 USD.

Nobels Proven Wrong?

It is important to remember that science very routinely corrects itself. The mistakes being corrected sometimes are because of misconduct, but other times it is because technologies or other understandings improve, and we learn that the initial conclusions were wrong. It is impossible to say which *really* happens more often, but I sense misconduct is less common. It should be no surprise that work that earned someone a Nobel Prize is no different. To date, no Nobel Prizes have been awarded for work later found to be fraudulent. Two examples of *disproven* works are Johannes Fibiger (1926, Physiology or Medicine) and

Enrico Fermi (1938, Physics). Fibiger was awarded the Nobel for the discovery of a parasitic cancer-causing worm. Although the worm without question is real, subsequent research proved there were no cancer-causing properties of this parasitic infection. Fibiger died (1928) before his work could be proven wrong.

Fermi, on the other hand, lived long enough to see his Nobel-winning work disproven and even agreed with the new (and correct) conclusions. Fermi had incorrectly concluded he generated new chemical elements during his nuclear chemistry work. He even went so far as to bestow names upon them. What Fermi had actually done, and Otto Hahn eventually figured out, was cause nuclear fission to occur. That this phenomenon was never documented before without question led to Fermi's incorrect conclusion. It simply could not—at the time—been the predicted outcome. Only later did additional work shed light on what was really happening.

Other award-winning works seem dubious, at best, with modern hindsight. Paul Hermann Müller (1948, Physiology or Medicine) was awarded the Nobel Prize for his discovery of the anti-insect properties of DDT, a substance that however effective at controlling mosquitos—and as a result malaria suppression—is now banned globally except in very exceptional circumstances. Its environmental impacts were later found to be severe enough that DDT is one of the central topics in Rachel Carson's *Silent Spring*, a work credited with igniting the environmental movement. Carlson died relatively young at 57 (in 1964). Perhaps she would have been a Nobel laureate (for Peace) for her conservation work had she not. Finally, Antonio Egas Monic (1949, Physiology or Medicine) earned his award for developing the medical procedure lobotomy, a procedure rarely used today.

Nazis and the Disappearing Nobel Prizes

Something you are unlikely to find in your standard world history text is that two Nobel Prizes were dissolved to keep them from being captured by the Nazis as they over-ran Europe. German scientists, Max von Laue (1914, Physics), for his discovery of diffraction of X-rays by crystals) and James Franck (1925, Physics), with Hertz, for their discovery of the laws governing the impact of an electron on an atom), sent their Nobel medals out of Germany to fellow physicist Niels Bohr in Copenhagen for safekeeping. If caught, von Laue and Franck likely would have been executed. Unfortunately, before long Copenhagen was likewise no longer safe; the Nazis were high-stepping through the streets. An associate of Bohr's, George de Hevesy, made the bold decision to hide the medals by *dissolving* them. That is right, he dissolved the gold medals much like any of us dissolve sugar in a cup coffee or a spot of tea. Gold is rather inert though, so dissolving it is difficult, requiring a solution that sounds like a bad idea to EVER make to the casual reader: a 3:1 mixture of hydrochloric acid:nitric acid. The result of this bold maneuver was two beakers of an orange solution. These beakers were set on a high shelf and de Hevesy eventually left the beakers behind as he fled for Sweden; as a Jewish scientist, he was not what you would call safe in Nazi-controlled land. After the defeat of the Nazis, De Hevesy returned to the lab. Miraculously (or perhaps it was the ghost of Alfred Nobel or maybe just luck), the beakers remained untouched. A little bit of chemistry later, the gold precipitated out and was sent to Stockholm. The Swedish Academy had the medals recast *from their gold*, and the medals were (re-)presented to their rightful owners in a 1952 ceremony. Personally, I find this story to be a wonderful thumb to the eye of arguably the most terrible scourge in human history poked by science.

Controversies and Snubs

Controversies

With honors such as the Nobel Prize, it should not be a surprise that there are occasional controversies, perceived snubs, or other general complaints and displeasure taken with both those who are awarded the prize and who are not. There has even been some wondering whether or not the Nobel Prize is good for science at all.[7] Although a complete treatment of these is not appropriate in this book series—it could likely be a book on its own—the interested reader is encouraged to perform a simple internet search for "Nobel Prize controversies." While the issue is far more commonly centered around the work or one of the awardees, entirely unrelated to the work of a Nobel Prize, in 2018, the Nobel Prize for Literature was delayed until 2019 because of sexual harassment claims made against one of the board members. This has to date been the exception, rather than the rule. A brief discussion of some of the biggest work-based controversies is appropriate here.

There have been several Nobel Prizes that have been criticized in the sense that many believe the awardee should not have received the award. The Peace Prize seems particularly vulnerable to this. Former U.S. President Barak Obama (2009, Peace) is one of the most controversial. This is primarily because he was given the award a mere nine months into his presidency. Nominations are collected *months* ahead of the actual award. This means that (then) President Obama, given the award "for his extraordinary efforts to strengthen international diplomacy and cooperation between peoples" was nominated incredibly early in his presidency; almost upon becoming president. The notion that he, or frankly anyone, could have done enough to be worthy of this award is almost laughable.

Another controversial Peace Prize was given to Yasser Arafat, who shared the award with then Israeli Prime Minister Yitzhak Rabin and Israeli Foreign Minister Shimon Peres (1994, Peace) because of their collective work on the

Oslo Peace Accords. Not only were these Peace accords unsuccessful at fostering peace—calling into question all three of the awards—Arafat was a key figure in many armed attacks in Israel; an awkward, at best, corollary.

Former U.S. Secretary of State Henry Kissinger's award (1973, Peace) was also very controversial. Recall, his Vietnamese counterpart Le Duc Tho refused to accept his share of the prize, citing that there was still conflict. There was also the bombing of Hanoi, ordered by Kissinger, during cease-fire negotiations but before the award was given.

Perhaps most scandalous of all is Fritz Haber (1918, Chemistry). Fritz Haber is the Haber of the Haber-Bosch process, a process still used largely unchanged today to make ammonia, essential to the commercial production of fertilizer. Because of this process, hundreds of millions, if not billions of people are fed. *This is not hyperbole*. Modern food production is possible because of his work. Whatever environmentalists want to (perhaps rightly) say about how sustainable this is and what the ceiling is, it is currently true. Just a few years *before* winning the prize, however, he orchestrated the first massive and deliberate chemical weapons attack during war, using chlorine gas against the allied forces in France during World War I.

D. Carleton Gajdusek (1976 Physiology or Medicine) is radically different from Fritz Haber. While Haber's crimes were done and well known *before* winning the prize, Gajdusek was not found to be a pedophile until years after his award. This means that Gajdusek's award was given ignorant of his vile actions while Haber was awarded his Nobel with full knowledge of the atrocities he committed. By no means am I comparing the crimes and ranking their *terribleness*. There is, in my opinion, a stark difference between *holding your nose and making a decision* (Haber) and *later wincing at your decision* (Gajdusek) as disgusting information comes to light.

Nobel Snubs

An extension of controversies is snubbing deserving winners. Before that, however, there are a number of giants of chemistry and science that could not have possibly been considered for the Nobel Prize, despite making contributions virtually unmatched in their importance. These pillars of chemistry, like those who solved the gas laws (Boyle, Charles, Gay-Lussac), Avogadro, and Lavoisier all predated Alfred Nobel's founding of the prize. In fact, some predated Nobel himself. Lavoisier is arguably the most impactful among them. In addition to demonstrating the importance of oxygen to combustion—helping to finally vanquish phlogiston—his law of conservation of mass is perhaps the single most important contribution to chemistry. Thereafter, chemistry was a quantitative, rather than qualitative endeavor. It became a science according to more modern interpretations of science. Unfortunately for chemistry and especially Lavoisier, he was executed by guillotine at 50 during the French revolution. In the more broad science sense, Charles Darwin died (much less violently than Lavoisier) in 1882, well before the creation of the Nobel Prize, rendering him ineligible. It should also be further noted that Darwin's work does not exactly fit into any of the areas recognized by the Nobel Prize. Physiology or Medicine is the closest match and would be quite the stretch.

Such early scientists notwithstanding, it should come as no surprise—go ahead and be disappointed though—that the selection of the winner of the Nobel Prize is subject to the flaws of human behavior. By this I mean that inevitably, biases, discrimination, and personal conflict have at least appeared to play a role in denying some otherwise worthy work being overlooked.

Even divorced from various biases, every year, it can be argued that "someone else" should have won the Nobel Prize, and this is probably true in every field, not just chemistry. In fact, this is the case for just about every award given across a very wide spectrum of recognition. Although sometimes the overlooked scientist eventually wins, the history of the Nobel Prize is littered with persons that it can be fairly asked, "How did *they* never win?" Chemistry is not without such controversies. Also, multiple women have been snubbed a Nobel Prize in various fields. These snubs are covered in the appropriate section of this introduction. Some examples of snubs include:

Gilbert Lewis could have been awarded the Nobel Prize in Chemistry for more than one important discovery. Perhaps his most noteworthy was the nature of a bond being the sharing of an electron pair. Lewis was nominated for the prize many times but was never awarded one. Reasons for him not winning one are difficult to prove, but there is some evidence that personal conflicts or less than fully informed opinions on the part of some evaluators playing a significant role.

Dimitri Mendeleev, father of the periodic table, was also never awarded a Nobel Prize in Chemistry. He allegedly came close, but the academy overseeing the award overruled the initial vote, changed the committee, subsequently holding a new vote that Mendeleev lost. Allegedly, this may have been at the behest of a rival (who at the time was a member of the Swedish Academy) of Mendeleev.

Henry Louis Le Châtelier was nominated for the prize but never awarded one despite his tremendously impactful contributions to chemical equilibria. It is possible that Le Châtelier's snub is a product of the historical fact that his most important work on equilibria was done too early (before the Nobel Prize was created), causing him to be overlooked because of the clause of "the previous year."

Chemistry and the sciences are not alone. Mahatma Gandhi, nominated five times for the Nobel Peace Prize, never won. Stephen Hawking, one of the most brilliant minds of our age was also never awarded a Nobel Prize for his work, nor was Carl Sagan. This may have some root in that Hawking's work could not be tested with contemporary tools, but maybe not.

Women and the Nobel

Over the years, across all the disciplines, the prize has been awarded to a woman 58 times, with one, Marie Curie, winning a total of two, in different fields (Physics with her husband Pierre in 1903 and Chemistry, alone, in 1911). The Physics prize has been awarded to women four times (2020, 2018, 1963, and 1903), Chemistry seven times (2020 (2 winners) 2018, 2009, 1964, 1935, and 1911), Physiology or Medicine 12 times (2015, 2014, 2009 (2 winners), 2008, 2004, 1995, 1998, 1986, 1983, 1977, and 1947), Literature 16 times (2020, 2018, 2015, 2013, 2009, 2007, 2004, 1996, 1993, 1991, 1966, 1945, 1938, 1928, 1926, and 1906), Peace, which leads the way with 18 times (2021, 2018, 2014, 2011 (three winners), 2004, 2003, 1997, 1992, 1991, 1982, 1979, 1976, 1946, 1931, and 1905), and the Economic Sciences prize has been awarded to a woman twice (2019, 2009). A numerical summary of these data is found in Table 0.2.

To date, all Nobel laureates have either outwardly presented as male or female. Even a casual perusal of the data should make clear that there is a wild imbalance with respect to the gender of the Nobel Prize winners. Currently, there has never been a winner who has announced being trans and none of them transitioned later in their lives after being awarded a Nobel Prize.

Though earlier on, the dearth of women recipients in especially the science fields could be explained away by a smaller number of women in the fields, compared to men, modern women win rates are nowhere near the averages reported by employers and (anecdotally, at least) observed by anyone paying attention to the world. It is difficult to pin down a plausible, data- or merit-based reason for this. Consequently, minds are left to wonder and inevitably wander to concluding discrimination and unfair judging. Although I want to say this is unjustified, I have no evidence to back up my stance. This is especially true for the awards with a more closed nomination process. Until and unless there is better accountability (perhaps through more transparent nominations) or systematic data-based rubrics that generate nominations or awards, this pall is likely to hang over the prizes.

Commentary on the dearth of women or person of color laureates, especially in the sciences, has been intensifying.[8] Although a shortage of nominees, at least for women has been cited,[9] concrete reasons for this lacking are far from agreed upon, a shortage of women in the fields is demonstrably false given that employment data indicate that the percentage of women working in the fields is higher than the percentage of women taking home Nobel Prizes. Even considering the unofficial transition to more of a lifetime achievement award, rather than greatest achievement of the last year award does not hold water as why women don't win more; women have been working in the sciences for decades. A list[10] has been compiled of deserving women chemists who arguably have been snubbed a Nobel Prize.

It would be quite easy to author an entire volume on a list of snubbed female Nobel-worthy scientists. The hardest part about such a list is determining when to stop. A short and nowhere near exhaustive list is:

TABLE 0.2 Analysis of Percent of Nobel Prizes Going to Women

AWARD	CHEMISTRY	PHYSIOLOGY AND MEDICINE	PHYSICS	PEACE	LITERATURE	ECONOMIC SCIENCES
Total	188	224	219	107	118	89
Women	7	12	4	18	16	2
% Women	4	5	2	17	14	2

Liese Meitner was one of the central contributors to nuclear fission, correcting Fermi's Nobel-winning work along with Otto Hahn. Despite being a long-time collaborator of Hahn's, including on his award-winning work, Meitner did not share his award (1944, Chemistry). It is exceedingly difficult for anyone to justifiably rationalize this oversight even if some of her work was hampered as she fled for her life (Meitner was Jewish) to Sweden with the rise of the Nazis to power.

Rosalind Franklin (one of the most infamous snubs) could fairly consider to actually be not snubbed in the sense that the award was given for the work she contributed to after her death. No sane and rational person can claim the committee waited for Franklin to die before recognizing the work. Nevertheless, it is impossible to ignore the fact that not one, not two, but three men were given the award related to her work. Part of what further enhances the ire of many is that particularly Watson and Crick collected little to no experimental results of their own regarding solving the structure of DNA, using especially Franklin's data on the way to do so. I for one believe—with admittedly no evidence—that she would have been awarded the prize had she not died. I see it happening in two different ways. First, Franklin could have been awarded the prize instead of Wilkins. This in my opinion is the most likely course as her data was viewed to be of higher quality than his. A second option would have been to make two different awards whereby Watson and Crick would be given one (likely Physiology or Medicine though perhaps Chemistry) and the other pair (who were-albeit adversarial-colleagues anyway) would be given the Physics or Chemistry prize.

Eleanor Roosevelt was an ardent advocate for civil rights and was never awarded the Peace Prize for her brave and noble work. Other civil rights activists (e.g., Rev. Dr. Martin Luther King, Jr.) were awarded the prize, so a claim that civil rights advocacy is insufficient to earn the award is weak, at best.

Joselyn Bell Burnell, who performed the work that led to the discovery of pulsars, sat by and watched while her colleagues won the Nobel Prize (1974, Physics). At the time, Burnell was a postdoctoral research student, further highlighting the disparity in credit between the Principal Investigators and those who work under their tutelage or mentoring.

Rachel Carlson's work is a cornerstone in the environmental movement. Had she not passed away in 1964, she likely would have been awarded a Nobel Prize in Peace for her impact on the environmental movement. Instead, it wasn't until 43 years later that Al Gore and the UN's Intergovernmental Panel on Climate Change were awarded the Nobel Prize for Peace for their work on increasing awareness regarding climate change.

Race and the Nobel Prize

The lack of racial diversity, particularly as it is measured by skin color, is as shocking as it is disappointing. Consider Table 0.3, which summarizes the data.

TABLE 0.3 Numerical Summary of Race and the Nobel Prize

PRIZE	CHEMISTRY	PHYSIOLOGY OR MEDICINE	PHYSICS	PEACE	LITERATURE	ECONOMIC SCIENCES
Total prizes	188	224	219	107	118	89
# Caucasian winners	173	214	195	80	102	86
% Caucasian	92	96	89	75	86	97
# Birth countries	38	38	35	47	49	19
# Affiliated countries	20	24	21	44	35	10

In most of the prizes, the numbers of Caucasian awardees are approximately the same level as women. Here, however, it may be easier to rationalize, but I want to fall well short of *justifying* it. This rationalization comes from the reality that at least in the cases of the research-based awards, the vast majority—in fact nearly all the winners—are affiliated with institutions within affluent countries that sometimes have billions of dollars annually invested in basic research. This higher level of funding inevitably leads to higher profile and higher quality—or at least higher sophistication—research. This is not to say that Caucasian researchers are for certain never overlooked. Once again, it appears inevitable that a black or brown person will be working at a high-profile and well-funded institution with vibrant research support. I even know more than a few such persons. As with women, there once again appears to be room to conclude only discrimination can produce this reality.

It is likely that an entire book can be written about the issue of general equality (be it gender, racial, or any other kind) in not just the Nobel Prizes but all manner of awards. One take that I heartily disagree with is the claim that it is a reflection of racism in the American education system.[11] Although the American education system is certainly rife with *inequality* that has many diverse roots, to claim that the American education system is at fault for a lack of diversity in a decidedly *international* award is absolutely off base. This may be part of why there is a dearth of racially diverse U.S.-based Nobel winners, it cannot possibly explain the issue on a global level. In fact, isolating the explanation to the education system in any country is potentially more counterproductive than it is helpful. Rather, I believe the cause to be more rooted in where there is economic power to invest in research.

The chemistry award is spread over 18 individual affiliation countries and twice was awarded to someone who had a professional affiliation with an institution in two different countries. By more than a factor of two, the U.S. leads the way with a whopping 82 winners. The chemistry award by birth country is a bit more diverse, covering 38 different countries of origin for the awardee. Again, the U.S. is the leader by a very wide margin. The dominance of the U.S. in the Chemistry Prize (and other awards) begins shortly after World War II. Prior to that, western Europe is the leader. A similar observation can be made (at least with respect to the rise of the dominance by the U.S.) in other fields as well.

Outlined in Tables 0.4–0.9 is a breakdown of country of origin and affiliation for all winners of each Nobel Prize category. The Physiology or Medicine award has been awarded to individuals affiliated with 24 different countries, hailing from 38 different countries. Once again, the U.S. dominates the field. As was seen with the birth countries of some of the Chemistry laureates, here we see countries that are no longer on the map. We also see how countries are referred to (e.g., Russian Empire vs. Russia and elsewhere U.S.S.R.) changing. This is deliberately done to try to provide the best possible historical context of the number of awards. What countries were known

TABLE 0.4   Country Breakdown by Affiliation and Birth for Chemistry Nobel Prize winners

CHEMISTRY

By affiliation

U.S.—82	Germany—33	U.K.—29	France—9	Switzerland—6
Japan—6	Sweden—5	Israel—4	Canada—3	Argentina—1
Austria—1	Belgium & U.S.—1	Czechoslovakia—1	Denmark—1	Finland—1
Italy—1	Netherlands—1	Norway—1	Switzerland & U.S.—1	
U.S.S.R.—1				

By birth

U.S.—55	U.K.—25	Germany—24	France—11	Japan—7
Austria–Hungary—6	Prussia—5	Canada—4	Netherlands—4	Sweden—4
Austria—3	Russia—3	Russian Empire—3	Scotland—3	Switzerland—3
British Mandate of Palestine—2		Egypt—2	Hungary—2	Norway—2
Palestine—2	Australia—1	Austrian Empire—1	Bavaria—1	Belgium—1
China—1	Denmark—1	India—1	Italy—1	Korea—1
Lithuania—1	Mexico—1	New Zealand—1	Poland—1	Romania—1
South Africa—1	Taiwan—1	Turkey—1	West Germany—1	

TABLE 0.5   Country Breakdown by Affiliation and Birth for Physiology or Medicine Nobel Prize Winners

PHYSIOLOGY OR MEDICINE

By affiliation

U.S.—115	U.K.—32	Germany—15	France—10	Switzerland—8
Sweden—7	Australia—4	Austria—4	Belgium—4	Denmark—4
Canada—3	Italy—3	Japan—3	Norway—2	Argentina—1
China—1	Dutch East Indies—1		Hungary—1	Netherlands—1
Portugal—1	Prussia—1	Russia—1	Spain—1	Tunisia—1

By birth

U.S.—80	U.K.—25	Germany—20	France—13	Australia—7
Sweden—6	Switzerland—6	Austria—5	Italy—5	Japan—5
Canada—4	Denmark—4	Austria-Hungary—3	Belgium—3	Netherlands—3
Russian Empire—3	Scotland—3	South Africa—3	Argentina—2	China—2
Norway—2	Poland—2	Spain—2	Austrian Empire—1	Brazil—1
Hungary—1	Iceland—1	India—1	Lebanon—1	Luxembourg—1
Mecklenburg—1	New Zealand—1	Portugal—1	Prussia—1	Romania—1
Russia—1	Venezuela—1	Wurtemberg—1		

TABLE 0.6   Country Breakdown by Affiliation and Birth for Physics Nobel Prize Winners

PHYSICS

Affiliation

U.S.—106	U.K.—27	Germany—20	France—13	Switzerland—9
Japan—7	Netherlands—7	U.S.S.R.—7	Sweden—4	Canada—3
Denmark—3	Italy—3	Russia—2	Australia—1	Austria—1
Belgium—1	China—1	Germany & U.S.—1	India—1	Ireland—1
U.S. & Japan—1				

By birth

U.S.—71	Germany—24	U.K.—23	Japan—12	Netherlands—9
France—8	Canada—6	Italy—6	Russia—6	Switzerland—6
China—5	Prussia—4	Sweden—4	West Germany—4	India—3
U.S.S.R.—3	Australia—2	Austria—Hungary—2	Austria—2	Denmark—2
Hungary—2	Poland—2	Russian Empire—2	Beglium—1	Czechoslovakia—1
French Algeria—1	Hesse—Kassel—1	Ireland—1	Luxembourg—1	Moroco—1
Norway—1	Russian Empire—1	Schleswig—1	Scotland—1	

TABLE 0.7 Country Breakdown by Affiliation and Birth for Peace Nobel Prize Winners

PEACE

Affiliation/residence

Organization—27	U.S.—21	U.K.—12	France—9	Sweden—5
Germany—4	South Africa—4	Switzerland—4	Belgium—3	India—3
Ireland—3	Argentina—2	Austria—2	Canada—2	East Timor—2
Egypt—2	Israel—2	Liberia—2	Norway—2	U.S.S.R.—2
Bangladesh—1	Burma—1	China—1	Colombia—1	Costa Rica—1
Democratic Republic of Congo—1	Denmark—1		Ethiopia—1	Finland—1
Ghana—1	Guatemala—1	Iran—1	Iraq—1	Italy—1
Japan—1	Kenya—1	Netherlands—1	Palestine—1	Philipenes—1
Poland—1	Russia—1	South Korea—1	Vietnam—1	Yemen—1

Birth

U.S.—19	France—8	U.K.—8	Germany—6	Sweden—5
South Africa—4	Switzerland—4	Belgium—3	Egypt—3	Ireland—3
Argentina—2	East Timor—2	Liberia—2	Norway—2	Poland—2
Russian Empire—2	Scotland—2	U.S.S.R.—2	Austria—1	Austrian Empire—1
British India—1	Burma—1	Canada—1	China—1	Colombia—1
Costa Rica—1	Denmark—1	Ethiopia—1	Ethiopian Congo—1	Finland—1
Gold Coast—1	Guatemala—1	India—1	Iran—1	Iraq—1
Japan—1	Kenya—1	Netherlands—1	Ottoman Empire—1	
Pakistan—1	Philipenes—1	Romania—1	Russia—1	South Korea—1
Tibet—1	Vietnam—1	Yemen—1		

TABLE 0.8 Country Breakdown by Affiliation and Birth for Literature Nobel Prize Winners

LITERATURE

By affiliation/residence

France—17	U.K.—13	U.S.—12	Switzerland—12	Germany—8
Sweden—8	Italy—6	Spain—5	Poland—4	Denmark—3
Ireland—3	Norway—3	U.S.S.R.—3	Austria—2	Chile—2
Greece—2	Japan—2	Mexico—2	South Africa—2	Belgium—1
Canada—1	Czechoslovakia—1	Egypt—1	Finland—1	Guatemala—1
Hungary—1	India—1	Israel—1	Nigeria—1	Peru & Spain—1
Poland & U.S.—1	Portugal—1	St. Lucia—1	Turkey—1	Yugoslavia—1

By birth

France—11	U.S.—10	Germany—7	Sweden—7	Spain—6
U.K.—6	Italy—5	Denmark—4	Ireland—4	Russia—4
Russian Empire—4	Japan—3	Poland—3	Austria—2	Austria—Hungary—2
Canada—2	Chile—2	China—2	India—2	Norway—2
South Africa—2	Belgium—1	Bosnia—1	Bulgaria—1	Crete—1
Colombia—1	Egypt—1	French Algeria—1	Guadeloupe Island—1	Nigeria—1
Hungary—1	Iceland—1	Madagascar—1	Mexico—1	Portugal—1
Ottoman Empire—1	Romania—1	Persia—1	Peru—1	Switzerland—1
Prussia—1	Trinidad and Tobago—1	Scheswig—1	St. Lucia—1	Tuscany—1
Tanzania—1	U.S.S.R.—1		Turkey—1	
Ukraine—1				

TABLE 0.9   Country Breakdown by Affiliation and Birth for Economic Sciences Award Winners

ECONOMIC SCIENCES				
By affiliation				
U.S.—68	U.K.—7	France—3	Germany—3	Norway—2
Sweden—2	Denmark—1	Finland – 1	Netherlands—1	U.S.S.R.—1
By birth				
U.S.—56	France—4	U.K.—4	Canada—3	Netherlands—3
Norway—3	India—2	Russia—2	Russian Empire—2	Sweden—2
Austria—1	British Mandate of Palestine—1		British West Indies—1	
Cyprus—1	Germany—1	Hungary—1	Israel—1	Palestine—1
Scotland—1				

as during the award gives readers insight into the timeline and geopolitical climate of the respective eras.

The award for Physics has been awarded to individuals affiliated with 19 different countries, with two awards being given to someone with multiple affiliations. As was the case in both the Chemistry and Physiology or Medicine awards, the number of birth countries ticks higher at 34. This is at least in part due to the ceasing to exist for some of these countries. Another part, however, is that people—including research scientists—relocate to other countries. In fact, after World War II, the United States "imported" (some may use the word poached) many topflight physicists from former enemy countries such as Germany.

Peace, the prize with the greatest diversity of all kinds, is one of the two award areas where the United States does not dominate. Part of why is that the leader is not any one individual but an organization. I have chosen to not try to parse out the country of origin of these organizations since it seems more important to me that the award went to an organization rather than an individual compared to that organization's country of origin. As far as individuals go, however, the U.S. retains its spot at the top, even if by a slimmer than used to margin. The award has been given to individuals affiliated with or living in 43 different countries at the time of the award with a birth country number of 47. Of the prizes considered so far—and Literature will have this phenomenon as well—Peace is different in that the winner is not always associated with some manner of academic or research institution, or other part of research or industrial science. The award for Peace has gone to pacifists, civil rights activists, religious figures, and politicians/heads of state.

The prize for literature is the second award area where the U.S. does not have dominance. Here, both France and the U.K. surpass the U.S.'s count, and Switzerland matches it. All told, affiliation countries ring in at 35 and birth countries at 39.

In Economic Sciences, the dominance flexed by the U.S. borders on obscene. A staggering 76% of the awards have been won by someone affiliated with the U.S. Affiliation countries ring in at a mere 10 and birth countries at 19. The smaller numbers of overall countries are no surprise since this award has existed for roughly half the time of the other awards.

Another trend is observed if one considers the prize statistics more carefully, and this trend too is observed across multiple disciplines as well, with exceptions being Literature and Peace. Most pronounced in chemistry, earlier in the history of the Nobel Prize, it was less common, though not unheard of, for there to be more than one awardee in any given year. The award going to more than one person is increasingly common. Whether this reflects an attempt to acknowledge more of the contributors, indecision, or a nod of sorts to the increasingly blended nature of sciences is difficult to pin down at this time, but it is a curious observation, nonetheless. Another possible explanation is that there is simply more science being done and with that increase in activity, there is a logical increase in the number of awards.

When considering a compilation of all the awards (Table 0.10), a total of 64 countries have had someone affiliated win the Nobel Prize with birth countries totaling 91 different countries. Currently, the U.S. has a healthy and likely uncatchable lead in both the affiliation number and birth number. It is worth noting, however, that there is a disparity of just over 100 (113 to be exact) in the affiliation number vs. birth number for U.S. awardees with birth trailing. This difference amounts to just over 25% of the U.S. awardees. This means that a healthy proportion of U.S.-based winners are immigrants of some kind.

What to do?

As it is virtually always the case that other work is at last as worthy as the awarded work, it is inevitable that this snubbed work was done by one or more women or someone with non-western origins. Without question, part of the problem is that these potentially deserving scientists are not even nominated. Since no contemporary list of nominations is published, it is impossible to know unless a committee member inappropriately leaks information or a nominator decides to disclose their nomination. Although causes for this potential lack of nomination are likely to be many, it is nearly impossible to exclude sexism or racism as one of them. We in the sciences may like to believe that we are above prejudices like racism, sexism, and all the other "isms" in our society, focusing our attention instead on independent interpretations of data; we in fact are not immune. Personal conflicts also inevitably rear their ugly heads. There are multiple pathways that may be effective at correcting this. For example, opening the nomination process more broadly or at least disclosing nominees may be low-hanging fruit for a more equitable

TABLE 0.10 Summary of Affiliation and Birth Country across All Awards All Time

ALL PRIZES

By affiliation

U.S.—404	U.K.—120	Germany—83	France—61	Switzerland—39
Sweden—31	Organization—27	Japan—19	Italy—14	U.S.S.R.—14
Denmark—13	Canada—12	Netherlands—11	Austria—10	Belgium—9
Ireland—7	Israel—7	South Africa—6	Spain—6	Australia—5
India—5	Poland—5	Argentina—4	Finland—4	Russia—4
China—3	Egypt—3	Chile—2	Czechoslovakia—2	East Timor—2
Greece—2	Guatemala—2	Hungary—2	Liberia—2	Mexico—2
Portugal—2	Bangladesh—1	Belgium & U.S.—1	Burma—1	Colombia—1
Costa Rica—1	Democratic Republic of Gongo—1	Dutch East Indies—1	Ethiopia—1	Germany & U.S.—1
Ghana—1	Iran—1	Iraq—1	Kenya—1	Nigeria—1
Palestine—1	Peru & Spain—1	Philipenes—1	Poland & U.S.—1	Prussia—1
South Korea—1	St. Lucia—1	Switzerland & U.S.—1	Tunisia—1	Turkey—1
U.S. & Japan—1	Vietnam—1	Yeman—1	Yugoslavia—1	

By birth

U.S.—291	U.K.—91	Germany—82	France—55	Sweden—28
Japan—28	Switzerland—20	Canada—20	Netherlands—20	Italy—17
Russia—17	Russian Empire—17	Austria—14	Austria—Hungary—13	
Denmark—12	Norway—12	China—11	Prussia—11	South Africa—10

(Continued)

TABLE 0.10 (CONTINUED)   Summary of Affiliation and Birth Country across All Awards All Time

ALL PRIZES

Australia—10	India—10	Poland—10	Scotland—10	Belgium—9
Ireland—8	Spain—8	Hungary—7	U.S.S.R.—6	Egypt—6
West Germany—5	Argentina—4	Palestine—3	Austrian Empire—3	
British Mandate of Palestine—3		Romania—3	Chile—2	East Timor—2
Liberia—2	Mexico—2	Portugal—2	Colombia—2	Turkey—2
Iceland—2	Luxembourg—2	New Zealand—2	French Algeria—2	Schleswig—2
Ottoman Empire—2		Israel—1	Finland—1	Czechoslovakia—1
Guatemala—1	Burma—1	Costa Rica—1	Ethiopia—1	Iran—1
Iraq—1	Kenya—1	Nigeria—1	Philipenes—1	South Korea—1
St. Lucia—1	Vietnam—1	Yemen—1	Bavaria—1	Korea—1
Lebanon—1	Mecklenburg—1	Taiwan—1	Venezuela—1	Württemberg—1
Hesse—Kassel—1	Morocco—1	British India—1	Ethiopian Congo—1	Gold Coast—1
Pakistan—1	Tibet—1	Bosnia—1	Bulgaria—1	Crete
Guadeloupe Island—1	Trinidad and Tobago—1	Madagascar—1	Persia—1	Peru—1
Tanzaniz—1	Cyprus—1		Tuscany—1	Ukraine—1
British West Indies—1				

process, assuming the cause is a lack of nominations for otherwise worthy awardees. This is because it would bring about at least a small measure of accountability or at least better transparency. As an academic, I like rubrics a lot. A well-designed rubric that generates nominations for the committee may better recognize the work of these snubbed scientists. Even the award process could be made more fair, consistent, and equitable using a rubric. The internet and data analysis allow for all sorts of metrics to be accessed and assessed. Metrics include the number of publications; the number of citations, including citations per publication; and so many others that can be found by anyone literate with an internet connection. Such an approach would be defensible, consistent, and blind to every attribute of the researcher except for the work itself. Although it is possible that such a rigid evaluation method is already practiced, the small number of women awardees casts doubt, serious doubt in my opinion, on this possibility. Another flaw that can be argued is that work that is infamous or debunked may at times get enough "attention" by way of *negative* citations that the rubric identifies it as worthy. This is where a human element would be able to (and ought to) override the rubrics. That said, opposition to such a rigid approach is not unfair. It can certainly be argued that such a rubric will disqualify a "dark horse" awardee. I at once disagree with this and find it irrelevant. First, a well-designed rubric can easily allow for such a nomination and award. Second, the Nobel Prize is not the NCAA Basketball Tournament, March Madness™. Nor is it the playoffs for some professional sport or even the Olympics where "Cinderella Stories" enthrall (and disappoint) millions. Although sometimes it takes years, decades even, to fully recognize and appreciate the enormous impact and importance, the Nobel Prize is not the place for a *feel-good underdog story*. It is for this reason that I believe a rubric that in some way evaluates the *already realized* impact of the works being nominated and considered is the fairest way forward. Quotas are one option that has been dismissed.[12]

The Chemistry Prize

The chemistry medal is the handiwork of Swedish sculptor and engraver Erik Lindberg. It represents Nature in the form of a goddess resembling Isis. In her arms, she holds a cornucopia and is emerging from the clouds, a veil covering her face, is held up by the Genius of Science.

Since its inception, through 2021, the chemistry award has been made 113 times. In 1916, 1917, 1919, 1924, 1933, 1940, 1941, and 1942, no award was given. One person, Frederik Sanger, was awarded the chemistry prize twice, though Linus Pauling and Marie Curie are chemistry laureates who have won two awards in different fields. Both of Pauling's awards (Chemistry and Peace) were unshared, and he remains the only *individual* to win more than one *unshared* Nobel Prize of any kind. Curie is one of two women to have earned an unshared chemistry award, the other being Dorothy Crowfoot Hodgkin in 1964 for her work on solving the structure of important biological substances using X-rays. Meanwhile, 60 men have taken home the prize solo. Once again, it is hard to justify this sort of disparity, even keeping in mind the growing number of instances where the award is given to more than one person. Curie remains the only woman to win more than one Nobel, one for Physics (1903) and one for Chemistry (1911).

Future of the Nobel Prize

There are no indications the Nobel Prize will cease to exist any time soon. Increasingly, some of the science-based awards are taking on more of a lifetime achievement award feel than one awarded for breakthroughs made in the preceding year. This does not take away from the prize. I cannot help but worry though that the Peace Prize, especially, will slowly but surely be increasingly accused of being political.

Increasingly controversial is who receives recognition for the work done vis-à-vis the Nobel Prize, and it is here I think some changes may come eventually. Although the Literature Prize and perhaps the Peace Prize are justifiably truly individual awards, scientific endeavors are more collaborative today than in Nobel's time. I would argue, in fact, that Nobel and his contemporaries could not have foreseen just how collaborative science would become. Not only are projects often conducted by teams of several researchers rather than an individual, the nature of any individual project is also increasingly collaborative since modern science is increasingly blended and interdisciplinary. As a result, it often demands multiple fields of expertise. This inevitably leads to a larger number of people. Projects also now take far longer to complete than 120 years ago, and the increase in time only further increases the number of people involved. This brings about a natural and exceedingly difficult to answer question of who among all these collaborators deserve(s) the award. Often, the PI, the Principal Investigator—the boss for the reader unfamiliar with the terminology—is the one who receives the award. This, even though in modern science this individual is very unlikely to have performed even one of the experiments behind the work. Why then do they get the award? It is as complicated as it is unfair.

First and foremost, this individual is *the constant*; they are the person who has been involved in the project from the beginning and is the only person (usually, anyway) who has had input to add to every aspect of the project since its inception. More for the inexperienced reader than the experienced reader, this individual is also the person who pays most of the (scientific) bills of the research through their grants. They also, whether they are in the lab doing actual scientific work or not, contribute heavily to the projects by way of suggestions for future/additional experiments and problem-solving. That said, there is some validity to complaints that such an approach is unfair and that it fails to appropriately acknowledge the work of the people performing the experiments in the laboratory. Unfortunately, I do not see any viable alternatives currently though it is for sure something to attempt to right. There has even been a suggestion that the Nobel Prize should be given for a topic, rather than to people.[13] How this relates to decidedly individual prizes such as Peace and Literature is unclear to me.

Future of the Chemistry Nobel Prize

A survey of what specifically the Nobel Prize in Chemistry is awarded for shows a clear rise in biochemistry (Table 0.11). Anecdotally, a colleague of mine has lamented that the Chemistry Nobel Prize seems to them to be more biology sometimes than chemistry. The Physiology or Medicine prize is also often as much biochemistry as it is Physiology or Medicine. Time will tell if this trend continues or shift to more energy-based science.

Predictions

If for no other reason but to have a little fun, I would like to make a few predictions on future Nobel Prizes. For soothsaying, I am going to restrict my actual predictions to the Physics, Chemistry, and Physiology or Medicine awards, though I will comment on the Peace Prize briefly as well. Regarding the Chemistry Nobel Prize, as the synthetic protocols currently referred to as C-H activation see further development, it is highly likely to earn someone a Nobel Prize. It is currently too early to identify a true frontrunner, but Melanie Sanford and M. Christina White are currently two leaders in the field. Alternatively, carbon sequestration or the conversion of carbon dioxide back to gasoline or other oil products are also important endeavors that would be the carbon and environmental equivalent to Haber's Nobel-winning work. Although it is without question worth asking, "why would we remake oil products from carbon dioxide?" it is important to recognize that the oil industry provides way more than just energy. It also provides starting materials and solvents for various critical materials such as textiles, dyes, and arguably most important of all, the pharmaceutical industry. An alternative source of these important substances or of fuels will be very important as these non-renewable resources dwindle in their supply and/or access due to geopolitical conflicts and increase in their cost.

Taking Physics next, if life—even simple life—were ever to be discovered elsewhere in the solar system (e.g., Mars), those responsible for doing the experiments or at least those who designed them will almost certainly win a Nobel Prize. Such a discovery would have incomprehensible ramifications for humankind. If recreational space travel becomes more popular, it is plausible (in my opinion, anyway) that either Elon Musk (*via* Space X) or Jeff Bezos (*via*

TABLE 0.11 Summary of Subdisciplines of Chemistry Receiving the Nobel Prize

YEAR	SUBDISCIPLINE	YEAR	SUBDISCIPLINE	YEAR	SUBDISCIPLINE	YEAR	SUBDISCIPLINE	YEAR	SUBDISCIPLINE
1901	P	02	O	03	G	04	G	05	O
06	G	07	B	08	N	09	G	10	O
11	N	12	O, O	13	I	14	E	15	B
16	None	17	None	18	I	19	None	20	P
21	N	22	A	23	A	24	None	25	G
26	G	27	B	28	B	29	B, B	30	B
31	G, G	32	I	33	None	34	N	35	N, N
36	A	37	B, B	38	B	39	B, O	40	None
41	None	42	None	43	P	44	N	45	B
46	B, B, B	47	O	48	A	49	P	50	O, O
51	N, N	52	A, A	53	I	54	G	55	B
56	P, P	57	B	58	B	59	A	60	EV
61	EV	62	B, B	63	M, M	64	A	65	O
66	G	67	A, A, A	68	P	69	G, G	70	B
71	G	72	B, B, B	73	O, O	74	P	75	O, O
76	I	77	P	78	B	79	O, O	80	B, B, B
81	O, O	82	A	83	I	84	O	85	A, A

(Continued)

TABLE 0.11 (CONTINUED) Summary of Subdisciplines of Chemistry Receiving the Nobel Prize

YEAR	SUBDISCIPLINE	YEAR	SUBDISCIPLINE	YEAR	SUBDISCIPLINE	YEAR	SUBDISCIPLINE	YEAR	SUBDISCIPLINE
86	P, P, P	87	O, O, O	88	P, P, P	89	B, B	90	O
91	A	92	P	93	B, B	94	O	95	EV, EV, EV
96	O, O, O	97	B, B, B	98	P, P, P	99	A	2000	M, M, M
01	O, O, O	02	A, A, A	03	B, B	04	B, B	05	O, O, O
06	B	07	M	08	B, B, B	09	B, B, B	10	O, O, O
11	I	12	B, B	13	G, G, G	14	A, A, A	15	B, B, B
16	G, G, G	17	B, B, B	18	B, B, B	19	EN, EN, EN	20	B, B
21	O, O								

Subdiscipline key: A—analytical chemistry/instrumentation; B—biochemistry; EG—Energy storage; EV—environmental; G—general chemistry; I—inorganic chemistry; M—materials; N—nuclear; O—organic chemistry; P—physical chemistry.

Blue Origin) may be in line for a Physics Nobel Prize. In fact, what those two have done already may be worthy of a Nobel Prize. The development of the technologies by an independent company rather than a governmental entity is utterly amazing. The development of the technologies each company uses is as sophisticated or more than anything any Nobel Prize has been previously awarded for. The automated and reusable rockets are technological wonders to be sure. The other thing they have potentially done is popularized something related to science. As of this writing, recreational space travel accessible to the masses is a long way off. However, the first steps have been taken because of these efforts. All that said, going to the moon did not earn anyone a Nobel Prize, so it is perhaps foolhardy to think this will.

Technology related to renewable energy also may eventually earn someone a Nobel Prize in Physics. However, improvements that would likely be needed to make that a reality would need to be significant, either in the reduction of cost, increase of yield/conversion, or both. Actually, depending on the precise nature of the technology, this sort of breakthrough may in fact be more worthy of a Chemistry Nobel Prize than a Physics one. It is shocking to me that energy storage has only been part of one Nobel Prize, the prize in 2019 for lithium-ion batteries, awarded to John B. Goodenough, M. Stanley Whittingham, and Akira Yoshino.

The electric or self-driving car both represent technological breakthroughs absolutely worthy of a Nobel Prize and both Elon Musk also has a hand in. Both have enormous advantages over current technologies and as each improve, we are increasingly likely to reap all those benefits and more.

For Physiology or Medicine, it is low-hanging fruit to predict that the m-RNA technology that led to the most effective of the COVID-19 vaccines will earn someone the Nobel Prize. This will be especially true if it is found that this technology is universally applicable to other vaccines as well, and I predict it will be. Cancer vaccines and even anticancer manmade viruses have been developed and if they prove successful would without question be at or near the front of the line for the award.

Finally, although not exactly a prediction, I must admit I am shocked that the Nobel Peace Prize has not yet been awarded to the Bill and Melinda Gates Foundation. The philanthropy done by this organization to date is tremendous and world-changing. Because of this generous work, hundreds of millions of people in some of the poorest and most disease-ridden parts of the world have access to medicine. The foundation has also done amazing work to improve education and access to technology in classrooms, globally. My only guess as to why no Nobel Prize has been awarded to the foundation is an attempt to send a message that you cannot "buy" a Nobel Peace Prize. Nothing other than that sort of message can explain to me why this foundation has not been bestowed this (in my opinion deserving) honor.

The Prize and Society

Shortly after the announcement of the 2021 Physics Prize, an opinion appeared on CNN claiming, "This Nobel Prize is a game-changer."[14] This reflects the opinion of the article's author that—since the Nobel Prize here is related to climate change—this should turn the tides of belief in favor of the validity of anthropogenic climate change. This, because of an implied vouching (my words) done by the committee for the science by making this award. In my opinion, this is untrue. Those denying anthropogenic climate change are already willfully ignoring the widely agreed-upon science. One more group of scientists "endorsing" it is very unlikely to perturb their stance. Additionally, there are many people who (apparently without jest) believe the world is flat and/or that the world is a mere several thousand years old. Both beliefs are held even when confronted with mountains (pun very much intended) of scientific evidence. In fact, I may argue the exact opposite stance about the societal impacts of the award. The deniers may in fact lose even more faith in the scientific establishment, viewing the award as something agenda-driven rather than science-driven. Such possibilities—turning the tides of belief, further entrenching against, or anything in between—should *never* be the goal of the award. The goal of the Nobel Prize never was, is not, and never will be to convince the throes of non-science persons to believe a scientific finding.

ABOUT THIS SERIES OF BOOKS

This series of books focuses on the Nobel Prizes in Chemistry that have contributed to the field of synthetic organic chemistry. Such a broad scope necessitates decisions that may appear arbitrary to the reader. I assure you, no slight is meant by any decisions to exclude certain awards, especially those in the next section. For certain, awards other than those covered in these volumes *involve* organic chemistry, even synthetic organic chemistry. Herein, I have chosen to omit any that only utilize or apply synthetic organic chemistry rather than build it up. A section covering "honorable mention" Nobels attempts to justify some of the omissions. In essence, I have tried to create a story of how the Nobel Prize in Chemistry made synthetic organic chemistry what it is today. Are such lines blurry? In a word, yes, and to be fair, even I think some of the awards I have chosen to include are borderline, at best. As the field continues to develop, I yield that some currently omitted studies may merit inclusion. If such a scenario comes about, I will address it with later editions, volumes, and/or both.

Currently, the volumes are delineated chronologically. Such organization was chosen to allow for the easier creation of both later updated editions and future volumes. Furthermore, a thematic organization would be difficult, if not impossible, to balance and adhere to parameters of logical and comparable volume sizes. Additionally, some years where the prize is awarded for multiple works, one may be more related to organic synthesis than the other. In this sort of case, the unrelated work will be mentioned but not receive a detailed review. The mention will be restricted to the bare minimum necessary to retain historical accuracy for the award.

Volume 1

1902 Fischer "in recognition of the extraordinary services he has rendered by his work on sugar and purine syntheses."

1910 Wallach "in recognition of his services to organic chemistry and the chemical industry by his pioneer work in the field of alicyclic compounds."

1912 Grignard "For the discovery of the so-called Grignard reagent, which in recent years has greatly advanced the progress of organic chemistry."

and Sabatier "for his method of hydrogenating organic compounds in the presence of finely disintegrated metals whereby the progress of organic chemistry has been greatly advanced in recent years."

1950 Diels and Alder "for their discovery and development of the diene synthesis."

1965 Woodward "For his outstanding achievements in the art of organic synthesis."

Volume 2

1979 Brown and Wittig "for their development of the use of boron (Brown) and phosphorus (Wittig)-containing compounds respectively into important reagents in organic synthesis."

1981 Fukui and Hoffmann "for their theories, developed independently, concerning the course of chemical reactions."

1990 Corey "for his development of the theory and methodology of organic synthesis."

Volume 3

1994 Olah "for his contribution to carbocation chemistry."

2001 Knowles & Noyori "for their development of catalytic asymmetric synthesis" and "for their work on chirally catalyzed hydrogenation reactions."

and Sharpless "for his work on chirally catalyzed oxidation reactions."
2005 Chauvin, Grubbs, and Schrock "for development of the metathesis method in organic synthesis."

Volume 4

2010 Heck, Negishi, Suzuki "for palladium-catalyzed cross couplings in organic synthesis."
2018 Frances H. Arnold "for the directed evolution of enzymes" (shared with **Smith and Winter** "for the phage display of peptides and antibodies," both of whom are not covered here).
2021 List, MacMillan "for the development of asymmetric organocatalysis."

Honorable mentions

As one might imagine, choosing who to include and, more importantly, *not* include in this sort of compilation is a challenging task. I dare (jokingly) quip that this decision is even more difficult than awarding a Nobel in the first place as I am assuming the position of omitting recognized achievement. Perhaps only to relieve a guilty conscience of sorts, I mention—and attempt to justify—here a handful or so of my most difficult decisions to omit in chronological order.

1905—Adolf von Baeyer "in recognition of his services to the advancement of organic chemistry and the chemical industry, through his work on organic dyes and hydroaromatic compounds." Arguably, this omission is the most egregious. His contributions to the field of organic chemistry are, without doubt, great and far-reaching. However, as far as their improving or building up the field of synthetic organic chemistry, not only is his primary contribution to this field (the Baeyer–Villiger reaction) not related to his Nobel, his award-winning work does not offer new and versatile synthetic methods or a changing of the way synthetic chemistry is conceived and performed. Perhaps future editions will include Baeyer, for now, I stand by my decision to omit.

1923—Fritz Pregl "for his invention of the method of microanalysis of organic substances" (for chemical composition). The importance of microanalysis of organic substances cannot be overstated. This work allowed for nothing less than the far easier determination of the molecular formula of chemical substances. Although running the experiment may not be, performing the calculations using data from this experiment are a standard part of many, if not all, college-level general chemistry classes. Without such data, identifying the chemical structure of compounds would be far more arduous. All that taken into consideration, this does not contribute to synthetic organic chemistry, so it is omitted here.

1947—Sir Robert Robinson "for his investigations on plant products of biological importance, especially the alkaloids." This work showed troponin alkaloids can be made from three simpler molecules. Omitting Robinson may even be more egregious than omitting Bayer. What Robison showed is (effectively) complicated molecules (at least the tropin alkaloids) could be made from much simpler building blocks. Had R.B. Woodward and later Corey not *totally* changed and expanded what was possible, it would be harder to omit Robinson. The outstanding work by Woodward and Corey, however, sets the bar so much higher that it is too difficult to include Robinson.

1956—Hinshelwood and Semenov "for their researches into the mechanism of chemical reactions." It is inarguable that a better understanding of chemical reaction mechanisms—how on an atomic level and/or molecular level molecules rearrange to go from starting materials to products for the less experienced reader—helped to drive the field of organic synthesis forward. However, as it is impossible to identify a specific way that this work built synthetic organic chemistry, it is omitted here.

1984—Robert Bruce Merrifield "for his development of methodology for chemical synthesis on a solid matrix." Merrifield's award is basically for polymer-bound peptide synthesis and, without question, has totally changed peptide synthesis. Without question, his contribution has revolutionized synthesis. However, to date, this has been primarily focused on peptide synthesis quite narrowly. If it were to ever be expanded to more general synthetic organic methods, it would be harder to justify omission. Some work has been done to apply it to general organic synthesis, but in my opinion, more must be done for a work such as this collection.

2018—Smith and Winter "for the phage display of peptides and antibodies" while including Arnold "for the directed evolution of enzymes." In short, Arnold's work has the potential to lead to enzymes being developed that will permit easier chemical transformations, perhaps—in a best-case scenario—leading to reactions that can compete with the selectivities and high yields observed with biological systems *in-vivo*. Meanwhile, the work of Smith and Winter has no relation to synthetic chemistry.

GENERAL SOURCES

https://www.nobelprize.org/, last checked 6/29/22
https://chemistry.as.miami.edu/_assets/pdf/murthy-group/gnl_jensen-2.pdf,   last
    checked 6/29/22

CITED REFERENCES

1. https://www.nobelprize.org/nomination/chemistry/, last checked 6/29/22
2. https://www.nobelprize.org/nomination/physics/, last checked 6/29/22
3. https://www.nobelprize.org/nomination/medicine/, last checked 6/29/22
4. https://www.nobelprize.org/nomination/literature/, last checked 6/29/22
5. https://www.nobelprize.org/nomination/peace/, last checked 6/29/22
6. https://www.nobelprize.org/nomination/economic-sciences/, last checked 6/29/22
7. https://www.chemistryworld.com/features/are-the-nobel-prizes-good-for-science/3009557.article, last checked 6/30/22
8. https://www.usnews.com/news/best-countries/articles/2020-10-01/the-nobel-prizes-have-a-diversity-problem-worse-than-the-scientific-fields-they-honor, last checked 6/30/22
9. https://www.science.org/content/article/one-reason-men-often-sweep-nobels-few-women-nominees, last checked 6/30/22
10. Borman, Stu. "Women Overlooked for Nobel Honors". *Chemical and Engineering News*, September 11th, 2017, 22–24.
11. https://www.popsci.com/racial-inequality-nobel-prize/, last checked 6/29/22
12. https://www.bbc.com/news/world-europe-58875152, last checked 6/30/22
13. https://massivesci.com/articles/nobel-prize-science-gender-physics/, last checked 6/30/22
14. https://www.cnn.com/2021/10/06/opinions/physics-nobel-climate-change-lincoln/index.html, last checked 6/29/22

Fischer

1

WHAT WAS THE DISCOVERY AND WHO MADE THE DISCOVERY?

The 1902 Nobel Prize in Chemistry was awarded to Hermann Emil Fischer "in recognition of the extraordinary services he has rendered by his work on sugar and purine syntheses." Fischer's accomplishments were wide in scope and breadth and had great depth. These synthetic milestones are but two of several enormous impacts.

It is difficult, in fact, to isolate the work of Emil Fischer to just the work related to the Nobel Prize. For example, the Fischer indole synthesis (Figure 1.1), discovered decades prior, was unrelated to the award. This reaction is still employed today. Given the aromatic nature of the indoles and that of the purines, it seems likely that one informed the other or perhaps each informed the other.

In Fischer's own words in his Nobel lecture, he would attempt to explain "what organic chemistry is capable of as the loyal ally of physiology with refined methods of analysis and synthesis." Fischer's work with carbohydrates demonstrated very clearly that organic chemistry and perhaps especially synthesis is a bonafide field of enormous power and potential. This, as much as any of his many synthetic contributions, is Fischer's most important contribution to the field. His connecting to physiology has also been proven true as anything there is in science. The overwhelming majority (over 99%) of therapeutic substances are organic compounds.

Perhaps an even longer-lasting impact can be seen in Fischer's carbohydrate (sugars) work. In part of his lecture, when introducing the carbohydrate work referring to artificial carbohydrates, he claims "the methods applied would be sufficient to produce hundreds more such substances." This turned out not to be hyperbole, with modern hindsight. Fischer also correctly notes that some sugars (e.g., glucose and galactose) have the same molecular formula.

Fischer was the first to convert glycerol into the corresponding 3-carbon sugar, coined by Fischer glycerose, by treatment of glycerol with dilute nitric acid. By the time of his Nobel lecture, Fischer had used the conditions

DOI: 10.1201/9781003006831-2

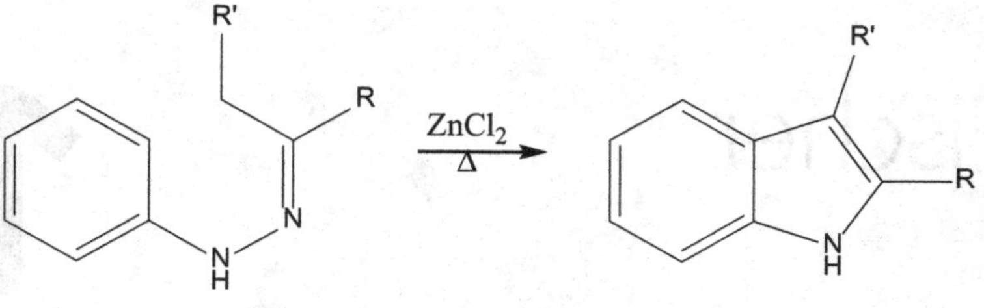

FIGURE 1.1:   Fischer indole synthesis.

pioneered by Kiliani (Figure 1.2) to add successive carbon atoms to sugars to generate carbohydrates with as many as nine carbon atoms. Fischer also proved without a doubt that there exists a two-carbon sugar molecule that behaves like all the other sugars, completing a quest to not just prepare larger sugars but also the simplest ones. In demonstrating this to be possible and that they all chemically behave the same way towards alkali and polymerize, it made clear that all these compounds belong to the same class of compounds. It is easy to overlook the importance of this simple-seeming task. It began to bring about order in the field. What Fischer contributed to, particularly with his preparation of many natural and synthetic sugars is that it is not Nature's exclusive province to produce optically active compounds, particularly the sugars. Fischer perhaps puts it best in his Nobel lecture freeing the preparation of optically active substances as "the privilege of the living organism." Fischer is accurate as he continues that "synthesis of both natural and artificial sugars provided a simple explanation for the chemical origin of this." He concludes, looking ahead to something modern chemistry frankly takes for granted to a great degree. He notes "It does not appear at all impossible to reproduce that asymmetrical synthesis artificially in the same way as it occurs in the natural formation of sugars."

To prepare the monosaccharides, Fischer used one of several reactions that now bear his name, the Kiliani–Fischer synthesis, whereby monosaccharides are increased in length by one carbon atom is an example. In this reaction, a shorter monosaccharide is treated with an aqueous cyanide salt, resulting in a diastereomeric mixture of cyanohydrins. Heating in water converts the cyanohydrin into the lactone, formed by nucleophilic attack on the nitrile by one of the secondary alcohols to make a six-membered lactone. Partial reduction with a sodium/mercury amalgam yields the diastereomeric monosaccharide mixture one carbon longer than the initial sugar. Recent improvements replace the conversion to the lactone and direct partial reduction of the nitrile to the imine, followed by hydrolysis to the new elongated monosaccharide.

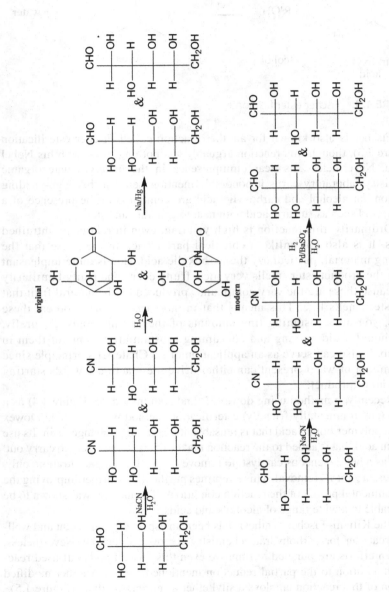

FIGURE 1.2: Kiliani–Fischer synthesis.

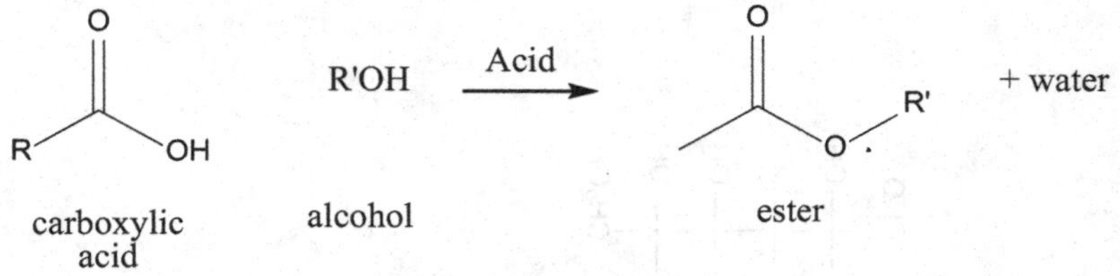

FIGURE 1.3: Fischer esterification.

Fischer is also known for another reaction called Fischer esterification (Figure 1.3), though this reaction arguably has nothing to do with his Nobel award. Nevertheless, its near omnipresence in the undergraduate organic chemistry laboratory earns it honorable mention here. In this high-yielding reaction, an alcohol and carboxylic acid are combined in the presence of a strong acid such as sulfuric acid with heat to generate an ester.

Ordinarily, this reaction is high yielding, even in relatively untrained hands. It is also not unlike a chemical parlor trick in the sense that the starting material, particularly the carboxylic acid, smells quite unpleasant while the product ester smells very nice, fruity even. The fruity familiarity is because these are the very compounds produced by the natural fruit that the ester smells like. This means that in most cases, when you eat these fruits, you are consuming tiny amounts of these compounds. Naturally, you should avoid making and consuming substantial amounts of them in the lab. It also can serve as an application of Le Châtelier's principle since the amount of water present can either drive the reaction towards starting material or product.

Recently, a method using dowex H⁺ and sodium iodide (Figure 1.4) as a green (environmentally friendly) esterification method was reported.[1] Dowex H⁺ is a polymer-bound acid that is reusable; it is a cation exchange resin. Its use over an acid that is added to the reaction makes the reaction easier to carry out. Also, as a heterogeneous catalyst, to remove the acid, a simple filtration, only, is necessary for its removal. This simplifies purification, further improving the environmental profile of the reaction conditions. The method was shown to be applicable to a wide range of alcohols and acids.

The Kiliani–Fischer synthesis has been an especially important and well-used reaction for carbohydrate chemistry for over a century. Nevertheless, modern efforts are pursued to improve even this ancient and well-used reaction in addition to the partial reduction mentioned earlier. Another modified version of this reaction employs a silyl ether as an intermediate (Figure 1.5).[2] These methods were developed to overcome moderate stability issues of the cyanohydrins under alkaline conditions. Although the substrate used in this

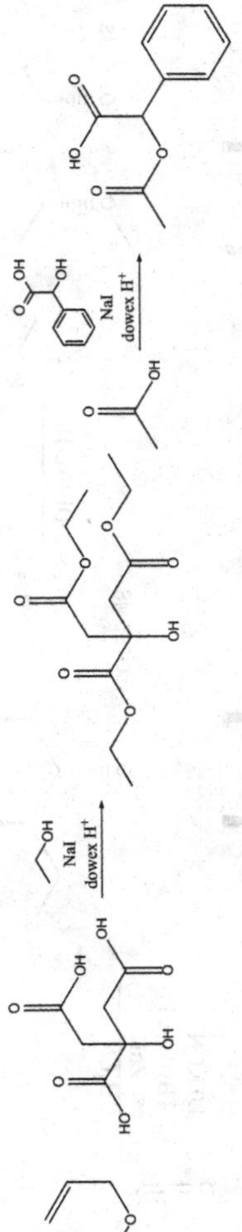

FIGURE 1.4: Fischer esterification reactions using dowex.

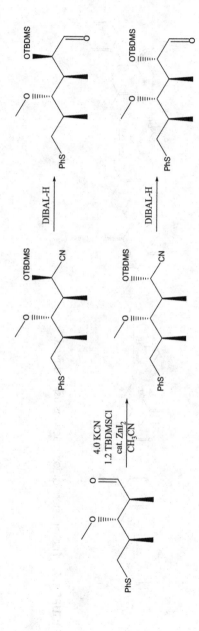

FIGURE 1.5:   Kiliani–Fischer synthesis-type reaction.

study was not, formally speaking, a carbohydrate, the chemical transformation here is the same as in the Kiliani–Fischer synthesis.

SUMMARY

In the grand scheme of things, more than for a particular chemical transformation or operation, Fischer's award was bestowed for a body of work contributing to the fields of purine and carbohydrate syntheses. The generalized nature of this award should not detract from the reality of the impact on the wider field of organic chemistry by both Fischer and the work. Though the impact may not be of the type that it enables the synthesis of a wide range of classes of compounds, as the first prize to be related to the field of organic chemistry and second prize overall, it legitimized the field, much like a politician yearns for endorsement from a predecessor or a wealthy donor. This impact cannot be overstated.

GENERAL SOURCES

https://www.nobelprize.org/uploads/2018/06/fischer-lecture-2.pdf, last checked 6/29/22

CITED REFERENCES

1. Turanen, P. A.; Leppänen, J.; Vepsäläinen, J. J. *ACS Omega*, 2019, *4*, 8974–8984.
2. Vogeleisen, A. F.; Uguen, D. *Tetrahedron Letters*, 1996, *37*, 5893–5896.

Wallach

2

In 1910, the Nobel Prize in Chemistry was awarded to Otto Wallach in recognition of his services to organic chemistry and the chemical industry by his pioneering work in the field of alicyclic compounds (essentially cyclic, nonaromatic). Similar to Fischer's award, Wallach's is not necessarily for an individual or set of chemical transformations. This second prize related to organic chemistry, much like the first several years earlier, is more for a breadth of work establishing synthesis as a field. Much of Wallach's work stems from these alicyclic compounds. Wallach's work, it should be noted, predates our modern understanding of aromaticity as a stabilizing force, though even by Wallach's time, its existence was recognized. Almost half a century earlier, Kekulé had deduced the structure of benzene. Although reading especially the earlier Nobel lectures, particularly the opening remarks, very much sounds like a hyperbolic love fest of ego-boosting, Wallach holds an especially humorous spot to me with his comment:

> The extraordinary honor of being allowed to address this illustrious assembly and to express my thanks for the superb recognition bestowed on my modest work by the Royal Swedish Academy of Sciences, gives me a feeling of happy pride at having been found worthy of distinction by such a distinguished body of men.

Why would this be humorous? The mention of a distinguished body of men, though no doubt refers to the Royal Swedish Academy of Sciences, is ironic in sense since the very next year, the same Academy will recognize the first woman to earn a Chemistry Nobel, the incomparable Marie (Madame) Curie, as of this writing was the first of two women to be awarded a solo Nobel Prize in Chemistry. By this time, recall, Marie had already shared a Physics Nobel Prize with her husband Pierre, so her winning a science Nobel Prize wasn't exactly groundbreaking.

   The long and the short of Wallach's work is that it centered around the essential oils whose core structure could not be traced back to benzene or one of its derivatives. These oils include the so-called turpentine oils: orange peel oil, caraway oil, peppermint oil, eucalyptus oil, fennel oil, thuja oil, and

DOI: 10.1201/9781003006831-3

camphor oil, just to name a few. These oils were initially a mystery as far as their natures were concerned.

There was initially a somewhat clumsy grouping of these oils, separated into terpenes and camphors, based on physical properties. Liquids—even at low temperatures—were called terpenes and solids were called camphors. Such a classification in hindsight may have delayed the full understanding of these compounds since a chemical reactivity-based grouping would allow for similarities to be found more readily.

A breakthrough came with the identification that most of these compounds (all of the important ones, according to Wallach) have ten carbon atoms. Moreover, the terpenes were found to share the molecular formula $C_{10}H_{16}$.

Wallach's work enabled the identification of distinguishing characteristics between the terpenes. Wallach proceeded to converting various oxygenated compounds into hydrocarbons and vice-versa. This allowed for elaborate and accurate connections to be drawn between the terpene essential oils. As some representative examples, in his Nobel lecture, he specifically calls out the conversion of pinene (ordinary terpene oil to Wallach) to carvone (caraway oil) with subsequent conversion of carvone to a mixture of limonene and terpinene. He also converted the ordinary terpene oil to eucalyptole (the main component of wormseed oil and oil from eucalyptus globulas) with terpineol (lilac-scented alcohol). This, in turn, was converted into l-limonene (laveo-limonene).

Wallach's goal was to find links between the various known compounds in the series to better understand their mutual relationships, and as such involved the progressive saturation with hydrogen of carvone, converting it into dihydrocarvone and tetrahydrocarvone, the latter initially only happening through the rearrangement of dihydrocarvone to carvenone first, which then gave way to hydrogenation but led to an optically inactive product. Only by using colloidal palladium as a catalyst could he prepare optically active tetrahydrocarvone. This synthetic protocol could, in turn, be applied to other syntheses as well, notably the other compounds typically unaffected by hydrogenation. Wallach also worked on a limonene series of compounds, obtaining a whole series of racemic mixtures from their optically active constituents by starting with either the laevo or dextro-limonene.

Working from prior knowledge that the compounds in this series were related to cymol (isopropyl p-methylbenzene), pinene, camphor, and other related compounds could be converted into cymol derivatives like carvacrol. Wallach correctly notes that as this family of compounds is no longer benzene-like are logically more reactive than their parent, whether by isomerization or addition.

After extensively talking about the chemistry he had done, Wallach goes on to wax somewhat poetic about the application to synthesizing essential oils. In particular, he attributes a better check on isolated essential oil samples made possible by their synthesis. These synthetic methods naturally gave rise to analytical methods of analysis as well, enabling the authentication of samples. He also notes how the terpenes were thereby found to be less important than the other compounds produced by the plant. Terpene-free essential oils followed and demanded a higher price tag. Wallach and his contemporaries had no way of knowing yet the role some terpenes play on attracting insects, often critical to pollination. He also attributes this growing synthetic power as the force behind the booming artificial-scent industry. Wallach goes on to borderline prophesy:

> We can now predict, with certain limitations, that synthetically producible compounds with a certain molecular structure will smell of peppermint, or camphor, or caraway seed or lilac, etc. As soon as we can wrest from Nature the secret of the internal structure of the compounds produced by her, chemical science can then even surpass Nature by producing compounds as variations of the natural ones, which the living cell is unable to construct.

Wallach doesn't stop there, also commenting on how well the synthetic power would enable the synthesis of other important industrial chemicals should global supply chains be disrupted.

Wallach's first major breakthrough came in identifying that the previously believed to be one substance was actually a mixture of the two enantiomers of limonene (Figure 2.1). Wallach's work picked up steam after a career move that amounted to a significant promotion.

By 1895, after the elucidation of the structure of α-terpineol (by Wallach and others at virtually the same time), Wallach's position at the top began to erode, but his mark had been left deeply enough to eventually earn him the Nobel. Early in the 1900s, Wallach began to author a collection of his works as a book, calling it "terpenes and camphors." Here, using terminology first used by Berzelius, he referred to terpenes as the ethereal (not the class of compounds, ethers) oils that resist solidifying at cold temperatures as terpenes and

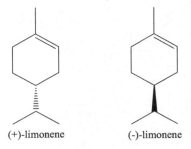

(+)-limonene   (-)-limonene

FIGURE 2.1: Chemical structure of limonene.

those that solidify as camphors. Part of what makes Wallach's work so important is that it placed on solid scientific bedrock the fragrance industry. He also set these classifications to be due to structural moieties, a far better organizational structure than one based on physical properties. Such organizational structures continue to this day. Wallach's contribution to the elucidation of the structures of the terpenes helped to conclude the triumph of structural theory, something that is integral to synthesis since it allowed insight into the chemical structure of the products. Wallach himself recognized the importance of his work to the fragrance industry.

SUMMARY

Without question, this chapter is very short on synthesis and chemical reactions. More than anything synthetic, Wallach brought order to the field of organic chemistry. Would that order have eventually come? Probably. The question of course is how long would that have taken without Wallach's work.

GENERAL SOURCES

Christmann, M. *Angewandte Chemie, International Edition*, 2010, *49*, 9580–9586.
https://www.nobelprize.org/uploads/2018/06/wallach-lecture.pdf, last checked 6/29/22
Partridge, W. S.; Schierz E. R. *Journal of Chemical Education*, 1947, *24*, 106–108.

Sabatier and Grignard

3

The 1912 Nobel Prize in Chemistry was split between Victor Grignard and Paul Sabatier. Grignard's contribution was to the development of organomagnesium compounds "the so-called Grignard reagent, which in recent years has greatly advanced the progress of organic chemistry." Sabatier, meanwhile, earned the prize "for his method of hydrogenating organic compounds in the presence of finely-disintegrated metals whereby the progress of organic chemistry has been greatly advanced in recent years." In short, both awards were given for specific synthetic methods. Even today, both methods are still ubiquitous and without question used daily worldwide.

Although it is obvious to any experienced organic chemist, to the novice or the student, the importance of the Grignard reagent can be easy to overlook. This importance lies in what these reagents do, specifically, that they facilitate the formation of new carbon–carbon bonds. This allows for the expansion of the size (i.e., the number of carbon atoms) of a molecule, permitting the formation of larger and ever more complicated chemical products from simpler starting materials.

Sabatier, on the other hand, developed what we now understand to be heterogeneous catalysts. Even Sabatier clearly recognized the catalytic nature of the reagents he employed. These catalysts allow for chemical transformations such as hydrogenation using materials that are easy, often trivial in fact, to remove from the product mixture. The catalysts explored by Sabatier were also remarkably effective. He even recognized the spoiling of some catalysts, something we now deliberately do to have the less reactive catalysts that only react with more reactive systems like alkynes.

After reviewing a brief history of hydrogenation methods in his lecture, Sabatier focuses particularly on the work of Moissan, who failed to correctly identify the major product of acetylene with freshly reduced nickel, cobalt, iron, or platinum black. Moissan had made ethane, not the presumed hydrogen, benzene, and other hydrocarbons. For Sabatier, it was a stroke of fortune that Moissan moved on to different work. Given my interest in scientific ethics, it is worth noting that Sabatier reached out to Moissan to verify he is no longer pursuing this as active research. Sabatier thought—and it turns out correctly—that the catalytic activity on porous platinum was not due to a local increase

DOI: 10.1201/9781003006831-4

in temperature due to condensation. Rather, he believed that a real chemical combination between the metal surface and surrounding gas was taking place. He also rightly attributed the reaction of acetylene to either an affinity between acetylene and the metal or the carbon or hydrogen of acetylene and the metal; the former now believed to be an accurate account of the mechanism.

Sabatier, along with Senderens performed tests on ethylene similar to those done by Moissan on acetylene. In this test, Sabatier and Senderens directed a stream of ethylene onto freshly reduced nickel, cobalt, or iron at 300°C. They observed the deposition of large quantities of carbon as the metal glowed intensely; the gas leaving the apparatus was confirmed to be ethane, not the hydrogen assumed by Moissan. In Sabatier's mind—and he is right—this could only arise if the metal induced hydrogenation of the ethylene. Furthermore, Sabatier confirmed the catalytic role of the metal by passing a mixture of ethylene and hydrogen down a column of nickel. In this experiment, the conversion of ethylene to ethane was found to proceed indefinitely.

During his lecture, Sabatier references a cruel bereavement in 1898 which made it impossible for him to do any useful work for many months. While internet searches have not revealed the source of this bereavement, it is refreshing to see one's mental health be taken so seriously, and we would do well to emulate that today. The apparent lionization of powering through at the expense of one's mental and emotional health is something we today seem obsessed with. That Sabatier was able to take nearly a year off and still come out well enough to go on to win a Nobel Prize should "give us all permission" to take a break when life goes awry.

After the success with ethylene, Sabatier and Senderens showed that the reaction with acetylene does in fact produce ethane, like their reaction with ethylene. They followed up with this study by showing that each reduced cobalt, iron, copper, and powdered platinum did so as well, even if less vigorously than nickel.

The pair then set out to use nickel catalysts to hydrogenate systems impervious to other conditions: the hydrogenation of benzene derivatives. To their delight, these reactions also worked, even at the relatively modest temperature of 180°C. Altogether, they were able to distill out the general statement of the method: "Vapour of the substance together with an excess of hydrogen is directed onto freshly reduced nickel and held at suitable temperatures (generally between 150 and 200°C)."

Senderens and Sabatier then went on to show the wide scope of the reaction, reducing both alkenes and alkynes to alkanes in very high yields and notably without byproducts or isomerizations. Nitriles gave way to amines under these conditions and aldehydes and ketones yielded alcohols primary alcohols from aldehydes, and secondary alcohols from ketones. Meanwhile, both carbon dioxide and carbon monoxide gave methane and an array of aromatic hydrocarbons (e.g., phenol, xylene, naphthalene) gave their fully hydrogen-saturated cyclic counterparts.

Sabatier and another coworker, Mailhe, continued, converting aromatic halogen derivatives to hydrocarbons; the hydrogenation of allyl products and unsaturated acids such that their function is retained; and the hydrogenation of oximes, amides, isocyanic esters, carbylamines, diketones, quinones, cresols, xylenols, and others. Even diphenols, pyroallol, and benzyl amine all surrendered to this transformation. Even the initially stubborn benzoic acid and its esters eventually yielded hydrogenated products.

Sabatier specifically notes that the purity of the metal is particularly important in this process, noting that traces of sulfur, bromine, or iodine all poison the catalyst and inhibit its activity. Today, we do this deliberately to mute the reactions and stop at otherwise reactive intermediates, making them the major product.

The specific group to be reduced was found to determine the purity of the nickel needed for the transformation. For example, ethylenic and nitro derivatives were more tolerant of impurities than aromatic compounds.

Curiously, Sabatier and Senderens noticed that the hydrogenation of acetylene proceeds better below 200°C; above this temperature the C–C bond breaks, generating a wide range of products with properties they remarked were like Pennsylvania petroleum. In the absence of excess hydrogen gas, condensing the product mixture yields a product they found to be like Baku petroleum. By modulating the equivalents of added hydrogen gas, still another petroleum-like product, this time resembling Rumanian petroleum was produced. Still other conditions (higher temperatures) furnished something like Galician petroleum.

All this talk of petroleum should seem weird or awkward today. They were simply relating their product mixtures to things they (and their contemporaries) were familiar with. In the intervening years, the chemical enterprise has developed far easier to understand and more generally applicable descriptions. Particularly awkward for the modern chemist should be focusing on the origin of the petroleum as part of the description. Such nomenclature trends that name things after where they have come from have fallen by the wayside, replaced by the far more formal IUPAC (International Union of Pure and Applied Chemistry) rules.

Other applications include the hydrogenation of liquid fatty acids; the conversion of acetone to isopropanol (which greatly reduced the latter's cost); the continuous conversion of nitrobenzene to aniline (which uses more poison-resistant copper rather than nickel to reduce the propensity to also hydrogenate the benzene ring); and the conversion of carbon monoxide to methane. Most valuable to Sabatier, though, is the hydrogenation of benzene derivatives to substituted cyclohexanes, previously only attainable by very laborious methods. Sabatier also hypothesized and proved that catalysts such as powdered copper could be used as dehydrogenation catalysts, converting alcohols to aldehydes and ketones at temperatures between 250 and 300°C.

Sabatier and Mailhe went on to use blue tungsten oxide and thorium oxide to dehydrate alcohols, inspired by the work using alumina by Ipatieff. That the catalyst for this process could be regenerated upon fouling supported Sabatier's mechanistic conjectures about the formation of temporary compounds, rather than some sort of heat released during condensation.

Since Sabatier's time, many metal catalysts have been developed. Some of them have even gone on to win Nobel Prizes. This includes the Palladium-catalyzed cross-coupling reactions (2010), covered in Volume 4 of this series, and the ring-closing metathesis catalysts (2005), covered in Volume 3 of this series. Although it is difficult to draw a straight line between Sabatier's work and modern metal catalysts, his early work laid the foundation for—arguably all—later catalytic chemical studies.

Stereoselective hydrogenation of ketones has even been demonstrated (Figure 3.1). Using a homogenous catalyst with an appropriate ligand, tremendous diastereoselectivities have been observed.[1]

This work was part of the 2001 Nobel Prize and is covered in greater detail in Volume 3 of this series. It can nevertheless be fairly considered an extension of the work done by Sabatier.

Although covered in more detail in Volume 3 with Knowles's and Noyori's Nobel Prize, some mention of stereoselective hydrogenation of alkenes is appropriate here as well. Parker, Hou, and Dong wrote an extensive perspective review[2] covering many of the different chiral auxiliaries used in tandem with active metal catalysts to bring about this powerfully selective reaction. That a reagent as small as H can be stereoselectively added to a planar substrate like an alkene is no small task and is of tremendous importance. In their perspective, the trio highlight four commercial chemicals (Figure 3.2)—levodopa,[3] a Parkinson's disease drug; pregabalin,[4,5] a drug for fibromyalgia; sitagliptin,[6,7] an anti-diabetic drug; and metolachlor[8] a herbicide—whose commercial productions have been impacted by these methods.

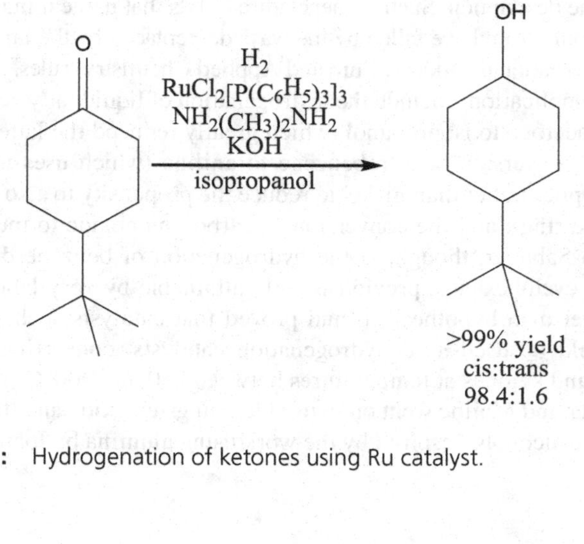

$$H_2$$
$$RuCl_2[P(C_6H_5)_3]_3$$
$$NH_2(CH_3)_2NH_2$$
$$KOH$$
isopropanol

>99% yield
cis:trans
98.4:1.6

FIGURE 3.1:   Hydrogenation of ketones using Ru catalyst.

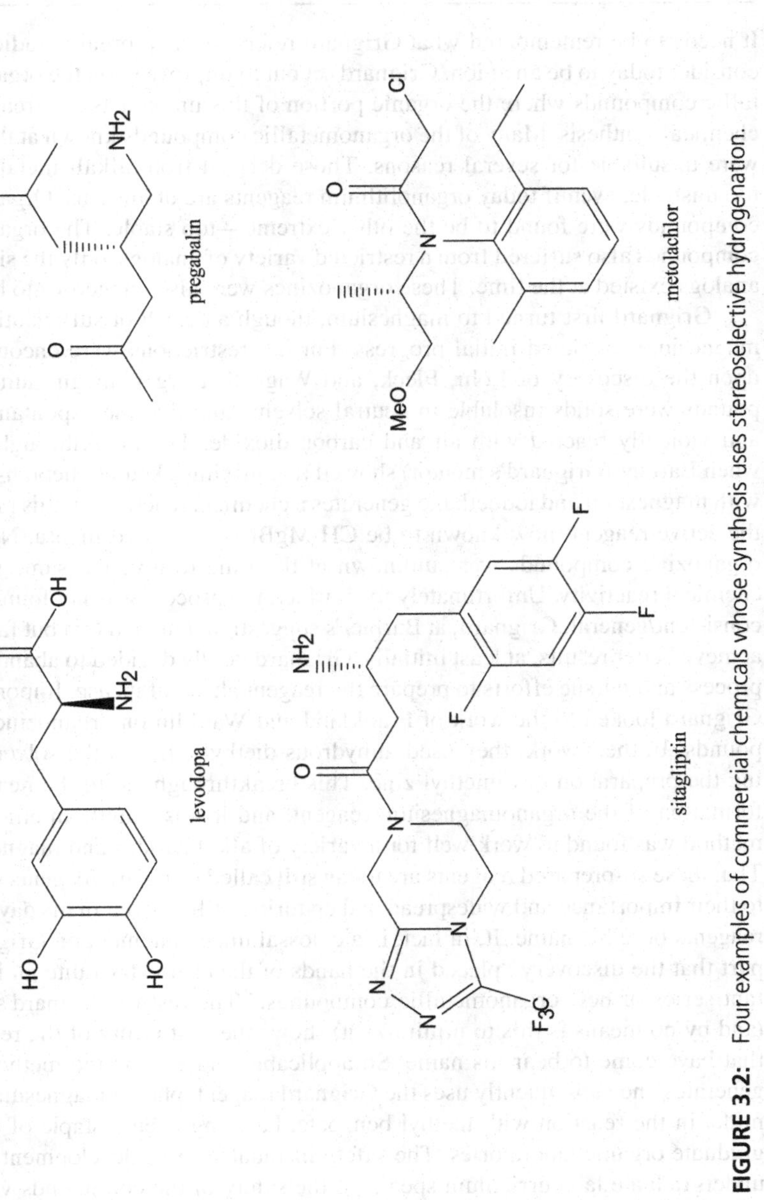

FIGURE 3.2: Four examples of commercial chemicals whose synthesis uses stereoselective hydrogenation.

GRIGNARD

It needs to be remembered what Grignard refers to as an organic radical, we consider today to be an anion. Grignard set out to improve upon the organometallic compounds where the organic portion of this union acts as a reagent in chemical synthesis. Many of the organometallic compounds known at the time were unsuitable for several reasons. Those derived from alkali metals were too unstable, even if today organolithium reagents are ubiquitous. Organozinc compounds were found to be the other extreme—too stable. The organozinc compounds also suffered from a restricted variety of analogs; only the simplest analogs existed at the time. These organozincs were also dangerous to handle.

Grignard first turned to magnesium, though a dearth of sufficiently pure magnesium restricted initial progress. Further restrictions were encountered upon the discovery of Lohr, Fleck, and Wagu that organomagnesium compounds were solids insoluble in neutral solvents and that they spontaneously and violently reacted with air and carbon dioxide. The breakthrough came when Barbier (Grignard's mentor) showed that mixing a ketone (heptan-2-one) with magnesium and iodoethane generates a chemical reaction. In this process, the active reagent, now known to be CH_2MgBr is generated in situ. Notably, organozinc compounds were unknown at the time to give the same sort of chemical reactivity. Unfortunately for Barbier, the process was not found to be consistent/general. Grignard, at Barbier's suggestion, pursued this but failed to achieve better results, at least initially. Grignard boldly decided to abandon the process and pursue efforts to prepare the reagent ahead of its use. Importantly, Grignard looked to the work of Frankland and Wanklin on organozinc compounds. In their work, they used anhydrous diethyl ether as the solvent during the preparation of dimethyl zinc. This breakthrough led to the near-total formation of the organomagnesium reagent, and it was soluble in ether. The method was found to work well for a variety of alkyl halides and magnesium. That these so-prepared reagents are today still called Grignard reagents speaks to their importance and widespread and enduring utility. Even in his day, these reagents bore his name. It, in fact, is a colossal understatement on Grignard's part that the discovery "placed in the hands of the chemistry quite an important series of new organometallic compounds." The rest of Grignard's work (and by no means is this to minimize it) shows the vast utility of the reagents that have come to bear his name. So applicable and easy is the method that generates and subsequently uses the Grignard reagent, phenyl magnesium bromide, in the reaction with methyl benzoate, has long been a staple of undergraduate organic laboratories. The safety insinuated in its development in the undergraduate lab curriculum speaks to the safety of the compounds via this method. Grignard also correctly identified the nature of the structure of these compounds as RMgX. The zinc compounds, in the meantime, have the formula ZnR_2. Part of how Grignard determined this is evidenced in the reaction of each with aldehydes with the organozinc compounds only one R group, only half of the alkyl group used to make the reagent, is added to the aldehyde while with the organomagnesium reagents, all the reagent is delivered.

Grignard also demonstrated the superiority of the organomagnesium reagent over the organozinc in their preparation. With the organomagnesium compounds, alkyl iodides, as well as bromides and chlorides can be used. The substitution level of the halide is also demonstrated by Grignard, showing that all substitution levels are viable partners. Even allylic halides, though troublesome in Grignard's time are viable partners. In the years since, vinylic and even aromatic halides have been shown to work very well. In fact, myriad such reagents are commercially available.

Grignard also showed that an alkyl organomagnesium compound could be used to deprotonate a terminal alkyne to generate an alkyne anion in what can effectively be considered a metal transfer reaction. He also showed the cyclopentadiene could likewise be deprotonated. Today, we know that this is because it is an aromatic anion.

Grignard also showed the formation of dimagnesium organic compounds, but this could only be generated if the halides are far enough apart in the starting material.

Grignard goes on to discuss the addition of these reagents to aldehydes, ketones, esters, and acid chlorides, though not all the research in these cases was performed by him and/or his coworkers.

Several advances have been made since Grignard's time, particularly related to the stereochemistry of the additions. Several models have been proposed to predict the direction of approach for any nucleophile towards carbonyls, including Grignard reagents. The most recent model is the so-called Felkin–Ahn model (Figure 3.3), which proposes that the nucleophile approaches the carbonyl in the conformation where the smallest group attached

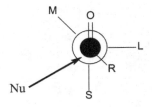

FIGURE 3.3: Felkin-Ahn model showing preferred direction of nucleophilic approach to carbonyls.

to the neighboring carbon atom is anti to the carbonyl with the largest group nearly perpendicular to it.

An impressively diastereoselective addition of ethyl magnesium bromide to a hydrazone was used as a key step in an improved synthesis[9] of the antifungal compound Noxafil® (Figure 3.4).

Diastereoselective 1,4 additions have also been shown[10] using copper additives (Figure 3.5). These conditions were designed to help access a key intermediate in the attempts at a synthesis of the macrolactone tylonolide, which is closely related to the macrolide antibiotic tylosin.

Substituted pyridine systems have also undergone stereoselective addition reactions. Here,[11] regioselectivity for the 4-position over the 2-position is also observed. In this series, the Grignard reagents show increasing selectivity for both diastereo- and regioselectivity, when compared to their organolithium counterparts (Figure 3.6). In Table 3.1, a brief survey of the results can be found.

Grignard additions have also been used in the key steps in the synthesis of natural products, for example, White's synthesis[12] of (±)-Euonyminol, a polyhydroxylated polycycle (Figure 3.7).

In their synthesis, a stereoselective addition to the β-carbon of an α, β-unsaturated ketone is performed (Figure 3.8). The initial deprotonation of the alcohol using LDA likely serves two purposes. First, since hydroxyl groups can protonate and thereby destroy a Grignard reagent, it likely has some protective function for the nucleophilic carbanion. The alkoxide also is argued to have a chelating effect with the magnesium of the Grignard reagent directing the addition to the same face of the molecule as the hydroxyl/alkoxide.

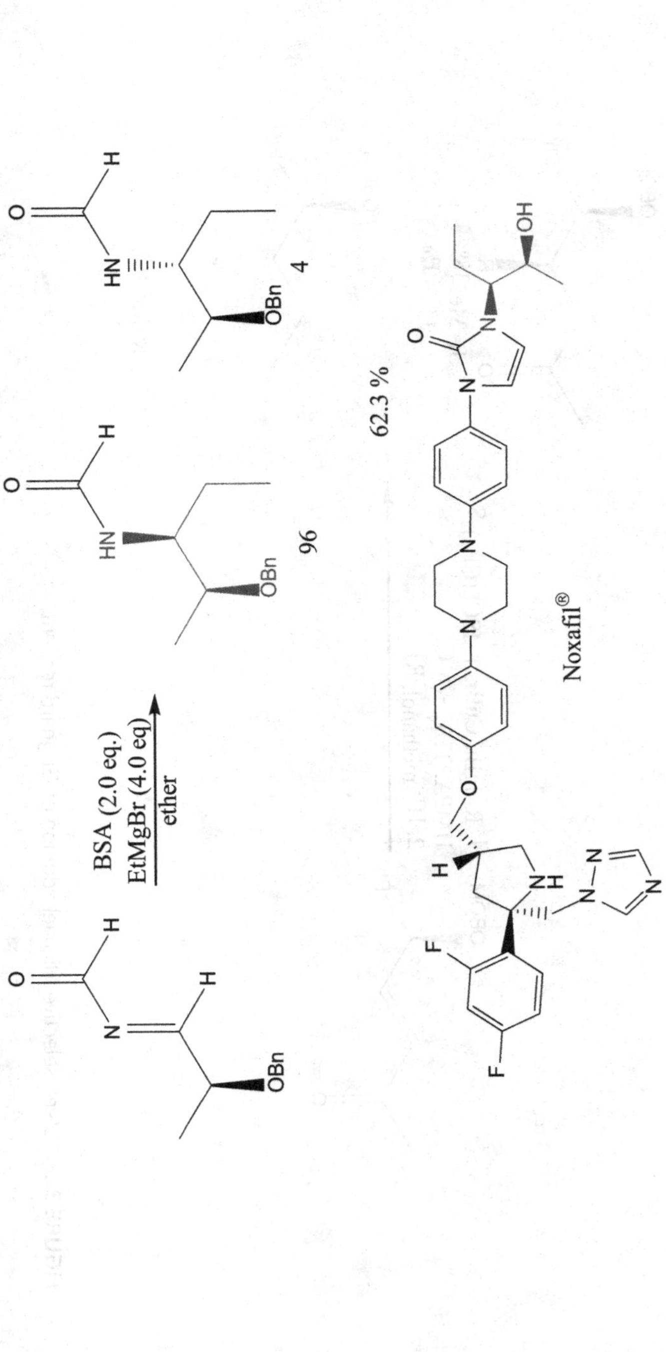

FIGURE 3.4: Use of stereoselective Grignard addition towards an antifungal compound.

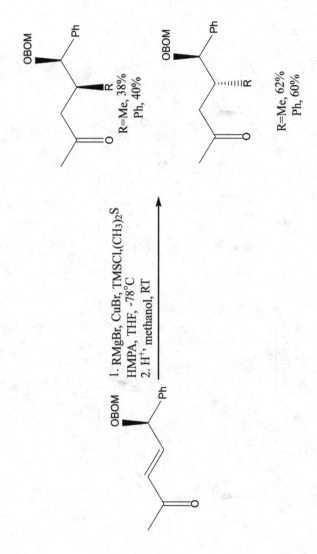

FIGURE 3.5: Stereoselective Michael addition of Grignard reagent.

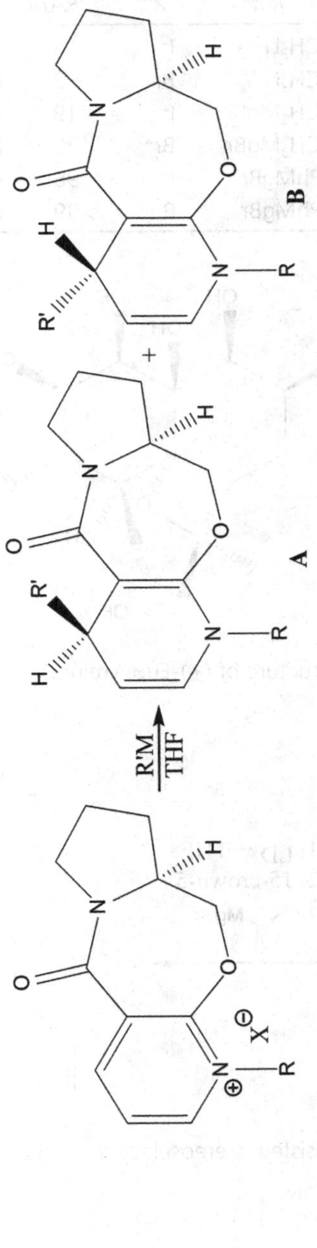

FIGURE 3.6: Stereoselective addition of Grignard and organolithium reagents to substituted pyridines.

TABLE 3.1 Comparison of Conditions and Ratio of Products

R	R'M	X⁻	RATIO A:B	
Me	CH_3Li	I⁻	3	1
Bn	CH_3Li	Br⁻	3	0.5
Me	CH_3MgBr	I⁻	19	0
Bn	CH_3MgBr	Br⁻	75	0
Me	PhMgBr	I⁻	98	0
Bn	PhMgBr	Br⁻	>99	0

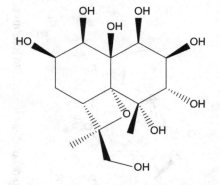

FIGURE 3.7: Chemical structure of (±)-Euonyminol.

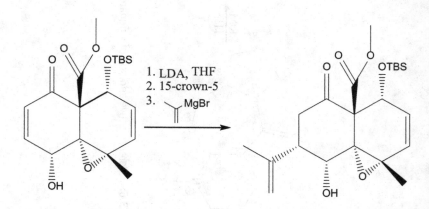

FIGURE 3.8: Alkoxide-assisted stereoselective Michael addition of Grignard reagent.

SUMMARY

With Sabatier and Grignard's award, the first fully synthetic organic chemistry Nobel Prize has been awarded. Over 100 years have passed since this chemistry was invented and the methods pioneered by these laureates are still used. In fact, I'd wager that at all times, in at least one lab, the synthetic methods described in this chapter are in use. They are that powerful and that common. Of the two, Grignard's methods are arguably the more important ones. This is because these methods allow for the incorporation of additional carbon atoms. That is, they allow for the building up of organic molecules and for the expansion of the molecular size. In future volumes, we'll encounter other such reactions; Grignard's was one of the earliest.

GENERAL SOURCES

https://www.nobelprize.org/uploads/2018/06/grignard-lecture.pdf, last checked 6/29/22
https://www.nobelprize.org/prizes/chemistry/1912/sabatier/lecture/, last checked 6/29/22

CITED REFERENCES

1. Noyori, R.; Ohkuma, T. *Pure Applied Chemistry*, 1999, *71*, 1493–1501.
2. Parker, P. D.; Hou, X.; Dong, V. M. *Journal of the American Chemical Society*, *143*, 6724–6745.
3. Lovering, F.; Bikker, J.; Humblet, C. *Journal of Medicinal Chemistry*, 2009, *52*, 6752–6756.
4. Hoge, G.; Wu, H.-P.; Kissel, W. S.; Pflum, D. A.; Greene, D. J.; Bao, J. *Journal of the American Chemical Society*, 2004, *126*, 5966–5967.
5. Schultz, C. S.; Krska, S. W. *Accounts of Chemical Research*, 2007, *40*, 1320–1326.
6. Hansen, K. B.; Hsiao, Y.; Xu, F.; Rivera, N.; Clausen, A.; Sun, Y.; Spindler, F.; Malan, C.; Grabowski, E. J. J.; Armstrong, J. D., III. *Journal of the American Chemical Society*, 2009, *131*, 8798–8804.
7. Savile, C. K.; Janey, J. M.; Mundorff, E. C.; Moore, J. C.; Tam, S.; Jarvis, W. R.; Colbeck, J. C.; Krebber, A.; Fleitz, F. J.; Brands, J.; Devine, P. N.; Huisman, G. W.; Hughes, G. *Journal of Science*, 2010, *329*, 305–309.

8. Dorta, R.; Broggini, D.; Stoop, R.; Rüegger, H.; Spindler, F.; Togni, A. *Chemistry a European Journal*, 2004, *10*, 267–278.
9. Saksena, A. K.; Girijavallabhan, V. M.; Wang, H.; Lovey, R. G.; Geunter, F.; Mergelsberg, I.; Mohinder, S. P. *Tetrahedron Letters*, 2004, *45*, 8249–8251.
10. Raczko, J. *Tetrahedron: Asymmetry*, 1997, *8*, 3821–3828.
11. Schultz, A. G.; Flood, L. *Journal of Organic Chemistry*, 1986, *51*, 838–841.
12. White, J. D.; Shin, H.; Kim, T.-S.; Cutshall, N. S. *Journal of the American Chemical Society*, 1997, *10*, 2404–2419.

Diels and Alder

4

The 1950 Nobel Prize in Chemistry was awarded to Otto Diels and Kurt Alder for "their discovery and development of the diene synthesis." Today, we know this reaction by another name—the Diels–Alder reaction, also called the Diels–Alder cycloaddition. This reaction, along with a few others, falls into a class of reactions called pericyclic reactions. In addition to cycloadditions, other examples of pericyclic reactions include electrocyclic, sigmatropic, and ene reactions. An in-depth discussion of these reactions is not entirely appropriate here, but later in this chapter, they will be expanded on to the extent that they are relevant to this chapter.

More so than Diels's Nobel lecture, which focuses on steroids, Alder's focuses on the reaction now known as the Diels–Alder reaction and other similar reactions, one of which has come to be known as the Alder-ene reaction, itself a pericyclic reaction.

To Alder's count, the Nobel he shares with Diels is just the second Nobel Prize in chemistry given to the field of organic chemistry. Although he is correct in the sense that it is the second award made for a specific reaction or transformation, as previously discussed, other awards also have relation to and contribute to the field of organic chemistry as a whole. Nevertheless, the diene synthesis—the Diels–Alder reaction (Figure 4.1)—is one of the most utilized reactions in the field and one of the most powerful. Work prior to Diels and Alder's showed that cyclopentadiene and quinone reacted with one another, but the structure of that product was not ascertained at the time (1906).

Diels and Alder had found the following general reaction to be the case for the reaction under study. In this reaction, a diene—specifically a 1,3 diene in the s-cis conformation—reacts with a dienophile (called a philodiene by Alder). The reaction proceeds to form a six-membered ring that overall has 2 less pi bonds than the combined starting materials. Regarding the stereochemistry of the product, at least as the starting materials are drawn here, substituent A and substituent F will appear on the same side as each other in 3-dimensional space in the product. Likewise, substituent B and substituent E will reside on the same side as each other in 3-dimensional space in the product. In the cases where the dienophile is a simple alkene, substituents G and I will be on the same side as each other in their 3-dimensional orientations

DOI: 10.1201/9781003006831-5

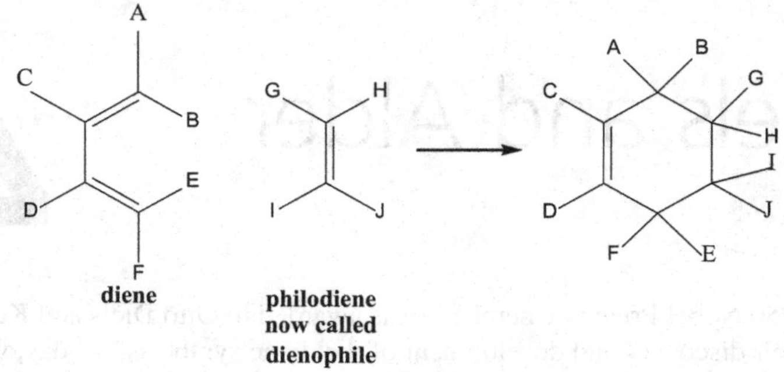

FIGURE 4.1: General diene synthesis (Diels–Alder reaction).

in the product and the same can be said for substituents H and J. Substituents C & D, meanwhile, residing on sp²-hybdized carbon atoms are trigonal planar in their configuration and thus have no 3-dimensional orientation traditionally drawn. As for how the relative positions in 3-dimensional space for substituents G/H and I/J vs. A/B and E/F, complicated factors to be discussed later determine the course of the reaction outcome.

It turns out that a very wide range of dienophiles is possible for this reaction. Alkynes, nitriles, and even some carbonyls are also viable partners with a diene in this reaction (Figure 4.2). Alder also discussed a variety of dienes that had been shown by that time to undergo this reaction. Although some aromatic rings do undergo a Diels–Alder reaction, the parent aromatic compound benzene, does not unless very specifically substituted as in styrene. Thiophene as well does not undergo this reaction and while pyrrole likewise does not, pyrrole does undergo a different reaction with dienophiles, to be briefly discussed later.

Even a casual glance at the species that undergo a Diels–Alder reaction (Figure 4.3) reveals that a variety of dienes and dienophiles will work. For example, the diene could be entirely acyclic, it could have both double bonds within a ring, it could have one double bond in a ring and the other attached to the ring. The reaction is also remarkably tolerant of functional groups, with the lactone shown being but one example. Likewise, the dienophile is very tolerant of a wide range of functional groups. The primary limiting factor in the functional groups is that functional groups with pi bonds themselves may also react as the dienophile rather than an alkene moiety in the dienophile. As noted by Alder in his Nobel address, dienophiles that possess an α, β-unsaturated carbonyl are particularly active in this reaction.

Of particular importance to Alder is maleic anhydride. Specifically, Alder notes that the switch towards using maleic anhydride rather than quinone

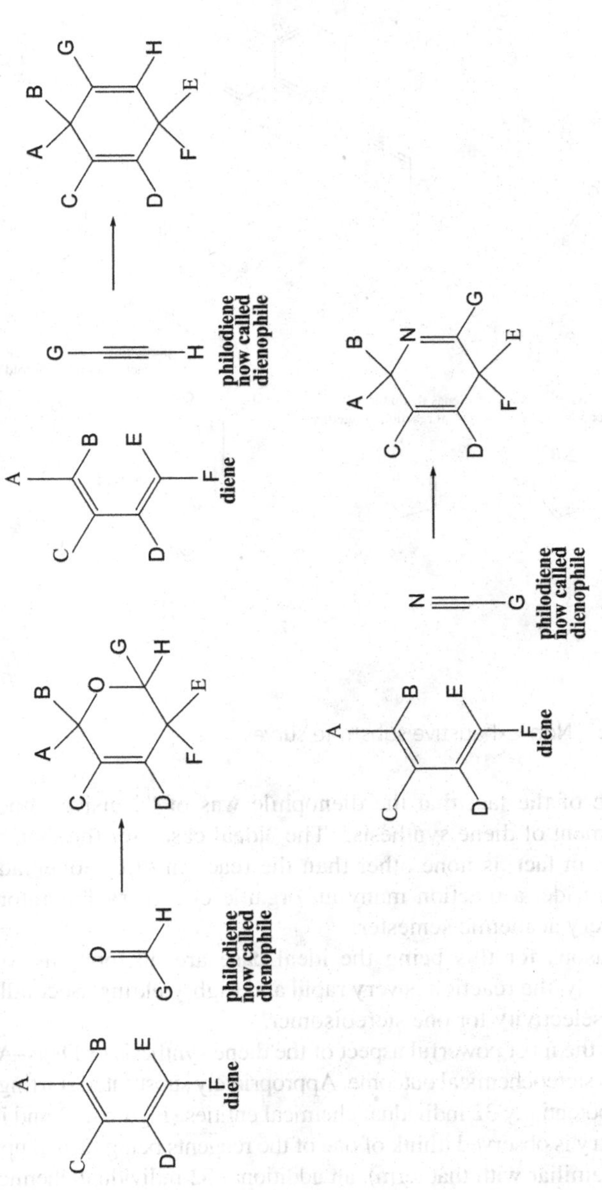

FIGURE 4.2: Varied dienophile diene syntheses.

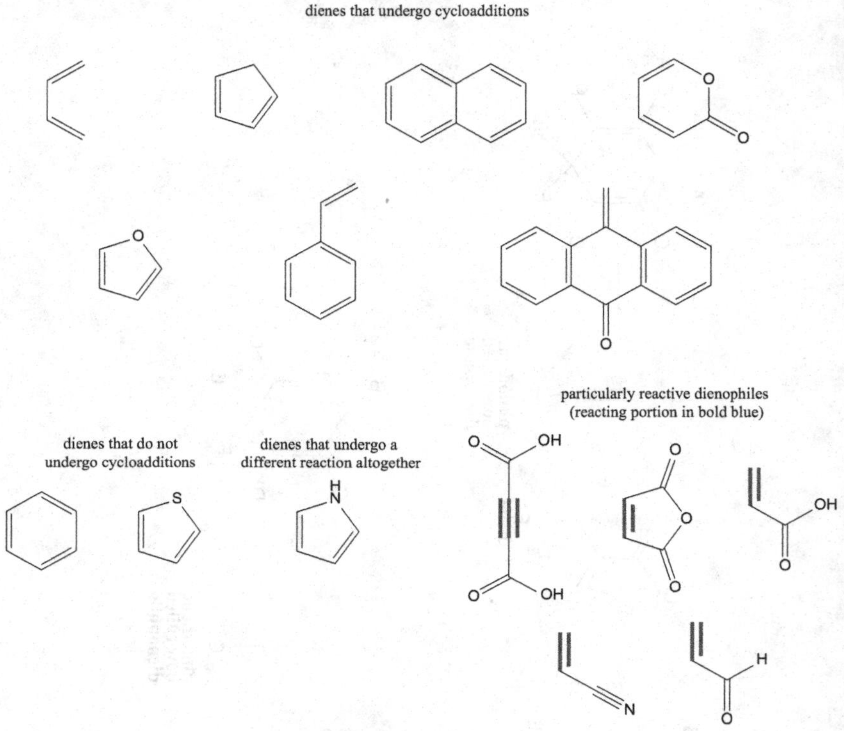

dienes that undergo cycloadditions

particularly reactive dienophiles
(reacting portion in bold blue)

dienes that do not
undergo cycloadditions

dienes that undergo a
different reaction altogether

FIGURE 4.3:   Non-exhaustive substrate survey.

was because of the fact that the dienophile was of "decisive importance in the development of diene synthesis." The "ideal case" for the diene synthesis (Figure 4.4), in fact, is none other than the reaction of cyclopentadiene with maleic anhydride, a reaction many an organic chemistry laboratory student performs every academic semester.

The reasons for this being the ideal case are not obvious to the non-chemist. Firstly, the reaction is very rapid and high-yielding. Secondly, there is remarkable selectivity for one stereoisomer.

Perhaps the most powerful aspect of the diene synthesis or Diels–Alder reaction lies in its stereochemical outcome. Appropriately substituted starting materials could yield potentially 32 individual chemical entities (Figure 4.5) and if different regiochemistry is observed (think of one of the reagents being turned upside down if you're unfamiliar with that term), an additional 32 individual chemical entities may arise. This means the reaction has the *potential* to lead to 64 different chemical entities. Despite this potential for chaos, the reaction reliably never gives more

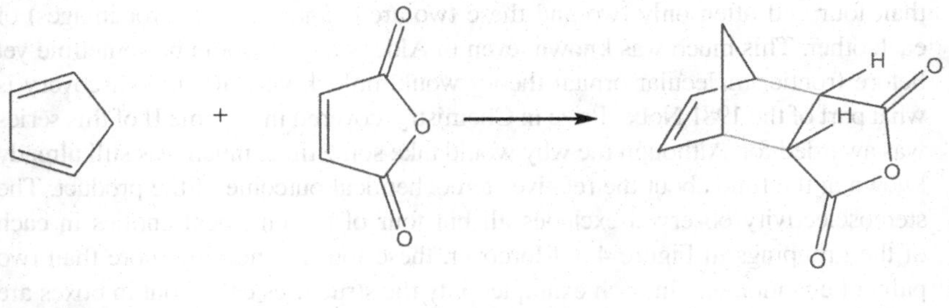

FIGURE 4.4: Alder's ideal diene synthesis.

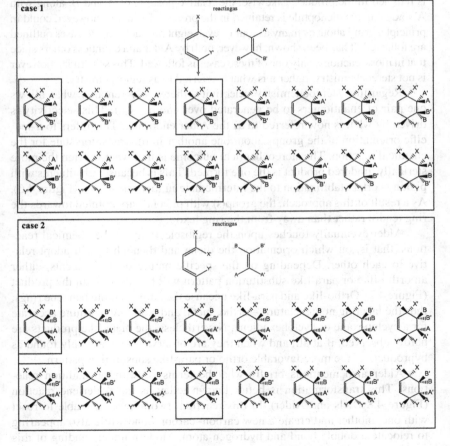

FIGURE 4.5: Possible products of diene synthesis reaction.

than four and often only two and these two are enantiomers (mirror images) of each other. This much was known, even in Alder's day. It would be sometime yet before frontier molecular orbital theory would unlock why and *this* discovery is what part of the 1981 Nobel Prize in Chemistry, covered in Volume II of this series was awarded for. Although the why would take some time, much was still already known at the time about the relative stereochemical outcome of the product. The stereoselectivity observed excludes all but four of the chemical entities in each of the groupings in Figure 4.5. Moreover, these four are nothing more than two pairs of enantiomers. In each example, only the structures called out in boxes are formed as products of these reactions. Often, however, only one of the vertically stacked pairs is observed. This relationship is described by saying that the relative position of X and Y, along with X' and Y' that exists in the starting material diene, is retained in the product. Likewise, the relative position of A and B, along with A' and B' in the dienophile is retained in the product. Even this, however, could in principle, bring about as many as four or even eight products if *both* cases outlined are followed. It has been shown, however, both by Alder and countless others since that in most reactions, only one of these cases is followed. This selectivity, however is not stereochemistry, rather it is what is referred to as regiochemistry.

Regarding stereochemistry selectivity, there only remains what allows one pair of enantiomers to be generated over another pair. This selectivity is driven by what is now referred to as the Alder endo rule. This refers to a specific orientation of the groups about one another in the transition state for the chemical reaction. The stereochemical outcome is that what is often the more sterically hindered product is the one formed. This is because substituents with pi bonds offer stabilization to the intermediate in this orientation (Figure 4.6). As a result of this approach, the group(s) with pi bonds are pointed towards the ring (endo) rather than away from the ring (exo)

Alder eventually touches upon the regioselectivity of the chemical reactions, that is, on which orientation the diene and dienophile will adopt relative to each other. Depending on the specific nature of the reagents, either an ortho-like or para-like substitution pattern will be observed in the product (Figure 4.7). Ortho-like and para-like because they are not quite benzene rings, thus the labeling nomenclature—ortho, meta, and para—aren't quite accurate for a cyclohexane or cyclohexadiene. Nevertheless, the meta-like products are hardly observed, if at all, and when they are observed, they are only found as byproducts of the more favorable ortho or para-like substitution pattern.

Alder also mentions briefly what he terms substituting addition reactions. This transformation is better known today as the Alder-ene reaction (Figure 4.8 Diels and Alder). In this reaction, two alkenes are able to react with one another and create a new carbon–carbon bond while also appearing to relocate a double bond and hydrogen atom. Modern understanding of this reaction suggests that the arrows shown in the image track where the electrons

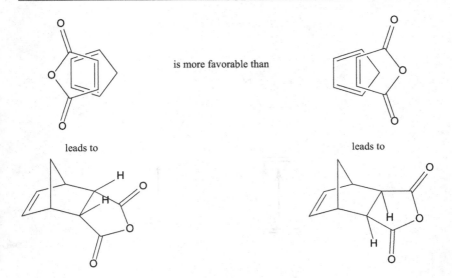

FIGURE 4.6:   Endo vs. exo selectivity of Diels–Alder reaction.

migrate during this chemical reaction, and such representations are commonly referred to as reaction mechanisms.

For the reader not familiar with such a representation, to interpret these arrows, we must keep a few things in mind. First and foremost, although individual steps occur in the order that they are presented, the individual arrows in any one step don't necessarily have to occur in the order of naturally reading right to left or top to bottom. In fact, a good many reactions, the ones shown here included, are what are known as concerted reactions. And so, in this case, we can take that entire merry-go-round or Ring Around the Rosie of electrons to happen all at once. So as the arrows are drawn here, the pair of electrons in the pi bond of the propene are donated to either of the alkene carbons of the maleic anhydride, resulting in the formation of the new carbon–carbon single bond. The pi bond simultaneously transfers its electrons to one of the hydrogen atoms on the CH_3 group of the propene. As this hydrogen atom is now receiving electrons, it no longer "needs" the electrons in the C-H bond, so it surrenders these electrons back to the carbon atom which in turn shares those electrons with the neighboring CH group. Notice that this CH group is the atom that would otherwise be losing electrons as the pi bond of the propene is donated to the maleic anhydride. It simply must get electrons back from somewhere and this is where.

We can likewise consider the mechanism of the Diels–Alder reaction (Figure 4.9), or if you prefer the diene synthesis.

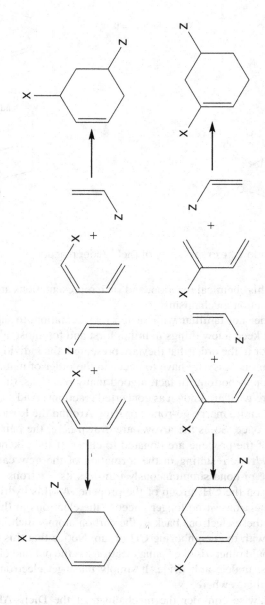

FIGURE 4.7: Regioselectivity of Diels–Alder reaction.

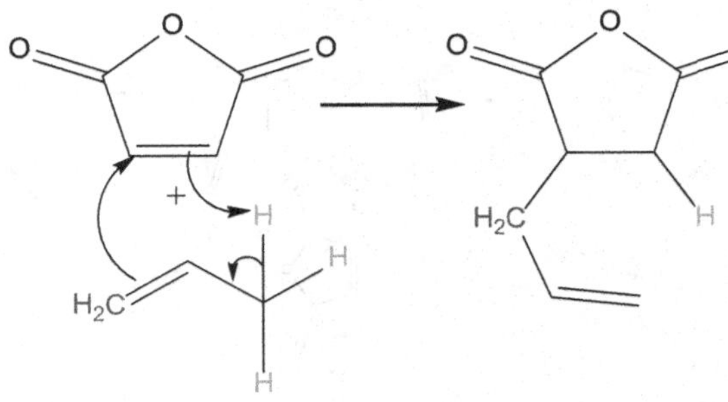

FIGURE 4.8: Substituting addition in the allyl position, a.k.a. the Alder-ene reaction, with reaction mechanism.

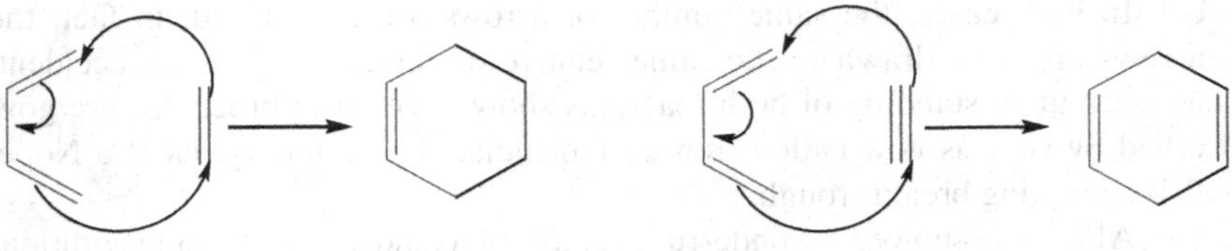

FIGURE 4.9: Diels–Alder (diene synthesis) reaction mechanism.

Notice the similarities here between this reaction mechanism and that of the Alder-ene reaction. Also, notice how the diene, according to the arrows, is losing one pi bond and relocating another, essentially, and the effect that this has on where the double bond is located in the product in the first case. Also, in the first case, the alkene dienophile is giving up its pi bond in order to make one of the new carbon–carbon bonds. It is for this reason that there is no pi bond in this position in the product. Compare this with what occurs in the case where an alkyne is the dienophile. Here, the diene portion is behaving identically to the first case. Likewise the dienophile is behaving exactly the same way in the second case. It may not seem like it is due to the presence of the second double bond, but if we keep careful track of what is occurring from the electrons' points of view, the same thing happens in both cases. Specifically, the dienophile gives up *one of its* pi bonds such that a new carbon–carbon bond may be formed. It also accepts a pi bond but not to function as a pi bond, but rather to create our second carbon–carbon bond. The only difference is that the alkyne, starting out with 2 pi bonds has one more after giving up one, while when the alkene gave up a pi bond, it gave up its only pi bond, leaving that hemisphere of the molecule without a pi bond.

Alder also notes a structural similarity between the diene synthesis reaction and substituting addition reaction. Reaction mechanisms as a concept were not yet fully developed in Alder's time, certainly not to the extent that they are today. Nevertheless, the similarity between the two is uncanny, and

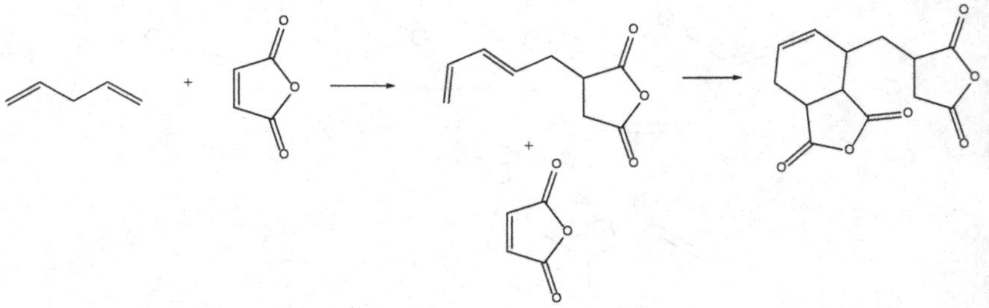

FIGURE 4.10:   Tandem Alder-ene and Diels–Alder sequence.

when the mechanisms are considered, they can be regarded as nearly identical. In both cases, the same number of arrows are drawn and, in fact, the arrows are even drawn in the same relative directions. This is no accident. Modern understanding of both reactions shows that these processes are governed by what is now called frontier molecular orbital theory, itself a Nobel Prize-winning breakthrough.

Alder also showed a wonderful example of a tandem substituting addition-diene synthesis example (Figure 4.10), or if you prefer the modern names, an Alder-ene reaction-Diels–Alder reaction tandem. In this case, pentane-1,4-diene was allowed to react with maleic anhydride. As the structure shows, while pentane-1,4-diene is a diene, it is not a diene capable of acting as a diene in a diene synthesis reaction. This reaction requires the diene to be conjugated. The diene in question, however, is perfectly suited to perform a substituting addition, or Alder-ene reaction with maleic anhydride. The resulting rearrangement of the pi bond results in a suitable conjugated diene which then reacts in the diene synthesis reaction. Of the two, the Alder-ene reaction was found to be slower. This means that the intermediate could not be isolated, only isolating the final product. As this ultimate product could only be formed if the first reaction occurs, the existence of the intermediate is inferred.

The Diels–Alder reaction has been employed in a vast number of natural product syntheses. The number is far too great to provide even a representative mixture. Some brief examples are discussed here. For instance, Danishefsky's synthesis of (±)-ipalbidine,[1] a non-addictive analgesic (Figure 4.11). Here, a hetero Diels–Alder reaction with a cyclic amine (an imino-Diels–Alder reaction) which occurs with desilylation and subsequent expulsion of an ethoxide with concurrent formation of the carbonyl.

Woodward also used the Diel–Alder reaction to kick off his famous synthesis of reserpine[2] (Figure 4.12). In this case, he used the reaction to generate the E ring of the natural product and also set three of the stereocenters for the target.

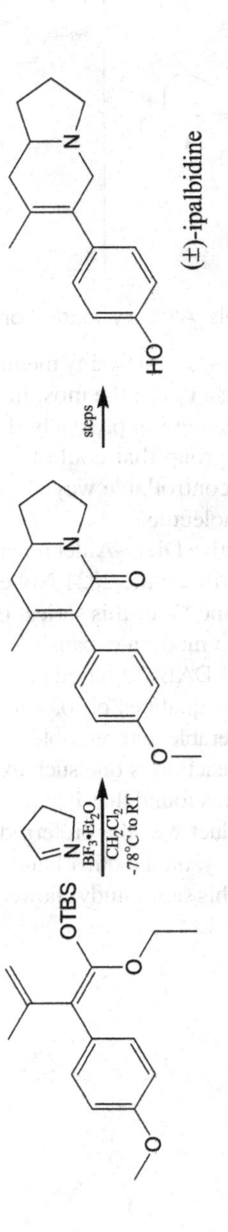

FIGURE 4.11: Use of Diels–Alder cycloaddition in the synthesis of (±)-ipalbidine.

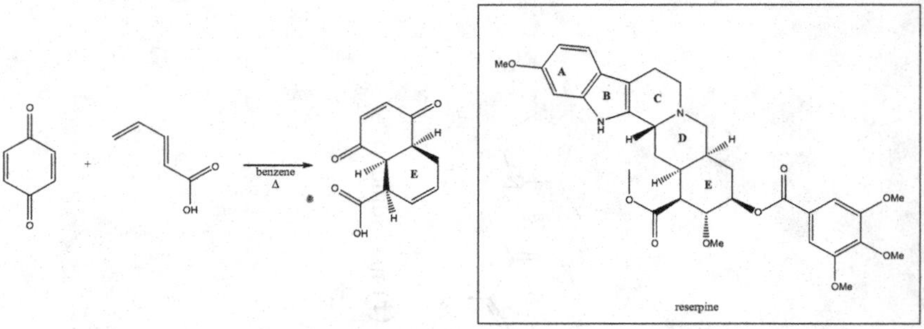

FIGURE 4.12: Use of the Diels–Alder cycloaddition in synthesis of reserpine.

The Diels–Alder reaction was not by any means the only reaction employed in this synthesis, it arguably isn't even the most important one. It did, however, generate a ring, three stereocenters in precisely the correct relative configuration and install a functional group that could be selectively transformed into additional stereocenters in a controllable way to further obtain desired stereochemistry elsewhere in the molecule.

A number of enantioselective Diels–Alder reactions have also been reported. Such reactions were associated with the 2021 Nobel Prize in chemistry and will see a deeper coverage in Volume IV of this series. Brief mention of a few examples is necessary, however, as a modern expansion of the original work.

One example is the use of DABCO-based chiral ionic liquids as an organocatalyst.[3] One of the important qualities of ionic liquids is that in a wide range of applications, they are recoverable and reusable solvents. This application to an enantioselective Diels–Alder reaction is one such example. Over a range of ionic liquids probed in the study, it was found that **1** (Figure 4.13) showed the best selectivity. As usual, the endo product was the preferred product, albeit by a meager 1.2:1 ratio. The enantioselectivity, on the other hand, was an impressive 87% ee in the case of the endo product. This same study showed that the exo product enjoyed the same levels of %ee (Figure 4.14).

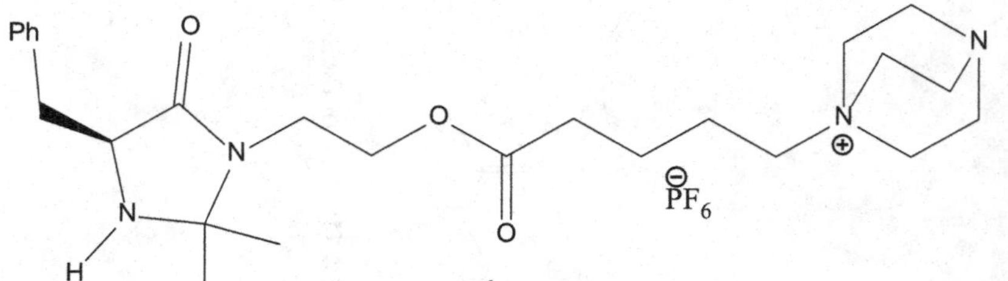

FIGURE 4.13: Ionic liquid catalyst for stereoselective Diels–Alder cycloadditions.

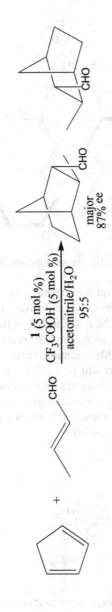

FIGURE 4.14: Stereoselecive Diels–Alder cycloaddition using a ionic liquid.

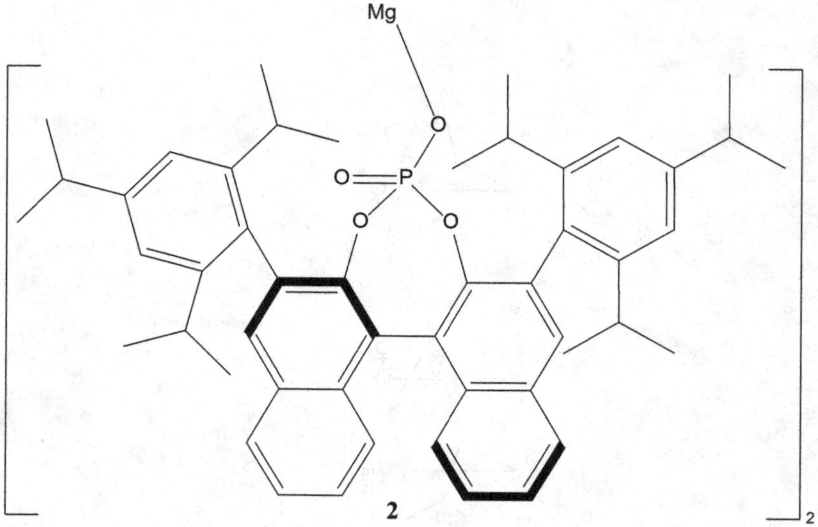

FIGURE 4.15:   Phosphonate salt for stereoselective Diels–Alder reactions.

Another example uses chiral magnesium phosphate complexes[4] such as **2** (Figure 4.15). A range of benzoquinone derivatives reacting with 3-vinylindoles was explored and with **2**, the best catalyst of their series, very impressive yields and %ees were observed over a range of reacting partners (Figure 4.16).

Another example of a highly enantioselective Diels–Alder reaction harnesses the hydrogen-bonding present in a chiral (and commercially available) alcohol to direct the enantioselectivity of a Diels–Alder cycloaddition.[5] Here, the chiral alcohol **3** (Figure 4.17) was found to provide the best enantioselectivity and product yield in the survey. Other derivatives replaced the 1-naphthyl ring with 2-naphthyl or phenyl (Figure 4.18).

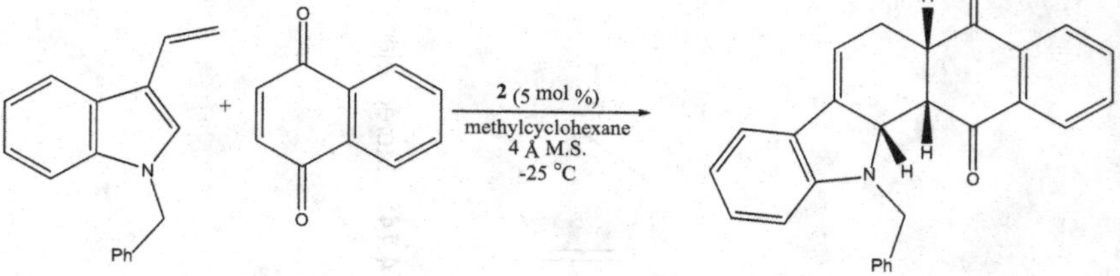

FIGURE 4.16:   Stereoselective Diels–Alder reaction using phosphonate salt.

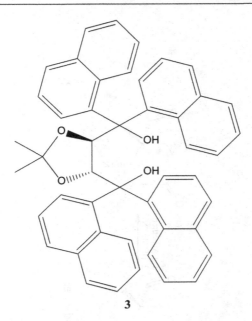

3

FIGURE 4.17:   Diol additive for stereoselective Diels–Alder reactions.

In this series, following the Diels–Alder reaction, the aldehyde was reduced with standard reducing agents and the silyl enol either is desilylated with HF to give the enolate which rapidly tautomerizes to the ketone.

Finally, even if MacMillan's work will see extensive coverage in a chapter of Volume IV, it simply must be mentioned here since this example represents the first enantioselective catalytic Diels–Alder reaction with *simple* α, β-unsaturated ketones as the dienophile. In this work,[6] hex-4-en-3-one and cyclopentadiene react in the presence of catalyst in a very high-yielding (89%) reaction with the expected endo selectivity (25:1 over exo and 90% ee Figure 4.19). The retention of endo selectivity across these reactions is important since it suggests the electronic effects underlying the normal course of the Diels–Alder reaction remain unchanged. This means we can more likely attribute the observed ee to some sort of steric effects brought about by the catalyst than some sort of electronic effect between the reacting partners. That is to say, these catalysts or chiral adjuvants do not appear to alter the molecular orbital interactions between the reacting partners.

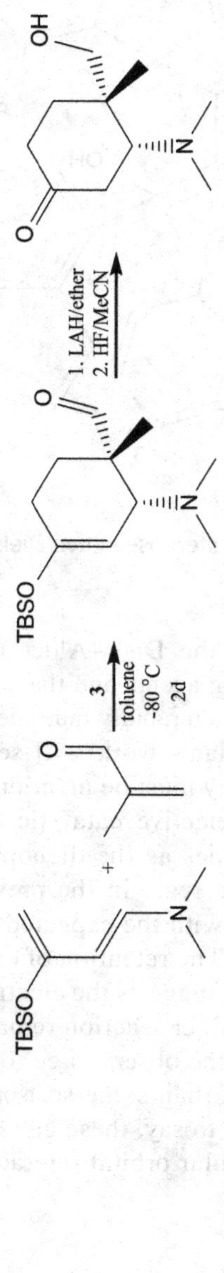

FIGURE 4.18: Stereoselective Diels–Alder cycloaddition using diol additive.

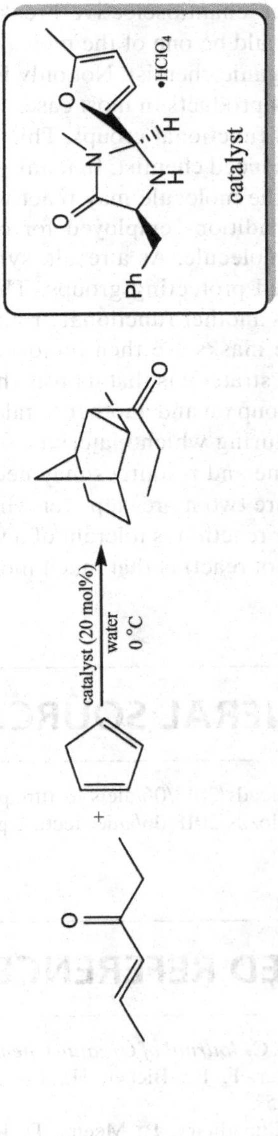

FIGURE 4.19: Stereoselective Diels–Alder cycloaddition using a chiral salt additive.

SUMMARY

Even in the absence of the enantioselective breakthroughs regarding the Diels–Alder reaction, it would be one of the most powerful reactions in the toolbox of the synthetic organic chemist. Not only is it remarkably selective (generating no more than 4 products in most cases out of a possible 64) it is also remarkably tolerant of functional groups. This last point is very easy to overlook for the less experienced chemist. In many multistep syntheses, certain functional groups in the molecule may react with one another or may react with the reaction conditions employed for other chemical transformations elsewhere in the molecule. As a result, synthetic chemists employ what has come to be called protecting groups. These groups temporarily mask a functional group as another functional group that does not have this troubling reactivity. These masks are then removed at the right time. The downside to the use of this strategy is that it costs the chemist two chemical reactions: one to put the group on and another to take it off. This means that there are two more steps during which material can be lost. It means there are two more steps that time and resources may need to be spent purifying a product. It means there are two more steps for which waste must be managed. That the Diels–Alder reaction is tolerant of a wide range of functional groups (meaning they do not react) is that much more empowering.

GENERAL SOURCES

https://www.nobelprize.org/uploads/2018/06/diels-lecture.pdf, last checked 6/29/22
https://www.nobelprize.org/uploads/2018/06/alder-lecture.pdf, last checked 6/29/22

CITED REFERENCES

1. Danishefsky, S. J.; Vogel, C. *Journal of Organic Chemistry*, 1986, *51*, 3915–3916.
2. Woodward, R. B.; Bader, F. E.; Bickel, H.; Frey, A. J.; Kierstead, R. W. *Tetrahedron*, 1958, *2*, 1–57.
3. Aalam, M. J.; Deepa; Chaudhary, P.; Meena, D. R.; Yadav, G. D.; Sing, S. *Chirality*, 2022, *34*, 134–136.

4. Bai, Y.; Yuan, J.; Hu, X.; Antilla, J. C. *Organic Letters*, 2019, *21*, 4549–4553.
5. Thadani, A. N.; Stankovic, A. R.; Rawal, V. H. *Proceedings of the National Academy of Sciences*, 2004, *101*, 5846–5850.
6. Northrup, A. B.; MacMillan, D. W. C. *Journal of the American Chemical Society*, 2002, *11*, 2458–2460.

Woodward 5

Our first volume ends with the chemist who was the forerunner of modern total organic synthesis and winner of the 1965 Nobel Prize in Chemistry, Robert Burns Woodward. Though without question, the field has made enormous strides beyond what even Woodward did, it is arguable that he started it all. Robert Burns Woodward was an enormously gifted and almost without parallel genius with regard to organic chemistry, particularly total synthesis. Only Woodward and E.J. Corey were awarded Nobels for what can broadly be described as genius. In the case of Woodward, his award was for "outstanding achievements in the art of organic synthesis." Corey's award (1990) is covered in Volume 2 of this series. In addition to his (Woodward's) work in the total synthesis of natural products, his work led to the correction of the structure of several natural products. The particulars of this are somewhat irrelevant here, but it should be understood by the less experienced reader that this means Woodward synthesized the structure that was reported by other researchers, but found this structure did not match the experimental data of the authentic natural product and so then prepared a compound with a (sometimes only slightly) different structure that *does* match the data of authentic sample, thereby determining (and correcting) the structure of the molecule. Thus, these corrections often mean that someone (in this case, Woodward) achieved an even more complicated synthesis than is done when the actual natural product is prepared.

The synthesis of strychnine, one of the most impressive of his many works, is believed by some to be what won him the Nobel Prize. Had his synthesis of quinine led to a source of it that was used by the Allied Forces during World War II to treat troops afflicted with malaria, it is possible that this synthesis would have gained better attention. However, the poor yield of the reactions reported by Woodward and Doering, the brevity of the quinine synthesis compared to strychnine, combined with the end of the war, makes this point moot. That said, as described later in this chapter, the synthesis of quinine is one of two controversies that have surrounded Woodward after his death.

Woodward was by no stretch a *one-trick pony*, nor was he a Luddite. He embraced new technologies, including spectroscopic techniques such as UV/VIS spectroscopy and infrared spectroscopy. His contributions to the

DOI: 10.1201/9781003006831-6

former led to what has come to be called the Woodward–Fieser rules. He also collaborated with Roald Hoffmann in the development of what is now called the Woodward–Hoffmann rules; these latter rules govern the course of pericyclic reactions, such as the Diels–Alder reaction (once called the diene synthesis).

An exhaustive comparison of Woodward's synthetic strategies with his contemporaries would fill volumes. Erik Sorensen compares Woodward to his contemporaries by pointing out that in Woodward's time, most synthetic chemists would start with compounds that had structures like their targets. These parts would then be linked together chemically using known chemical transformations/conditions. There is, in truth, nothing inherently illogical with this synthetic approach. Even today, modern synthetic chemists will take this approach if the starting material is cheaply available and plentiful in supply. The conditional availability though can be very restrictive, and Woodward changed that with his innovative strategies. Sorensen attributes Woodward's creativity to his mechanism-based reasoning.

Furthermore, even a half-hearted perusal of Woodward's publication record reveals that he did not invest effort into creating synthetic methods that are then applied to a variety of targets or a range of products. Such methodological development is now commonplace, even among those pursuing the total synthesis of complex natural products. Thomas Maimone comments on this, saying that Woodward considered the uniqueness of each target which helped dictate the overall strategies he would use. Woodward's synthesis of reserpine (discussed more fully below) is one of the works often pointed to as a highlight of Woodward's genius.

Descriptions of Woodward by his peers and students would make you think that you were reading about a dancer or some sort of artist, rather than an organic chemist. K.C. Nicolaou, for example, says that Woodward was characterized by "creativity, artistry, and elegance," with his synthetic endeavors as having a molecular complexity that were a quantum leap when compared to what was reached at the time by others.[1] Meanwhile, chemists such as Richmond Sarpong, who have only heard of Woodward as a historical figure, refer to him as "an institution."[1] His Nobel Lecture, other lectures, and many of his papers all read more like art than written word about an intense field such as synthetic organic chemistry.

In the time of Woodward's award, 1965, it is important to note that organic synthesis was a far newer field. Many of the modern characterization, purification, and synthetic techniques that are now commonplace did not yet exist or were in their infancy in Woodward's time. That Woodward was able to lead teams that completed such monumental syntheses is extremely impressive.

To rigorously cover the body of Woodward's work could fill a volume—multiple volumes, actually—on its own (See Table 5.1 for a brief list). Here, two

TABLE 5.1   Partial List of Woodward Syntheses

COMPOUND NAME	YEAR REPORTED	COMPOUND NAME	YEAR REPORTED
Cephalosporin C	1966	Illudalic acid	1977
Vitamin B$_{12}$	1972	Illudacetalic acid	1977
Chlorophyll	1960	Isolongistrobine*	1973
Tetracycline	1962	Tribromosantonin	1962
Marasmic acid	1976	Cholesterol	1951
The Penems	1979	dl-6-demethyl-6-deoxytetracycline	1968
Illudinine	1977	Cortisone	1951
N-methylisomorphinane	1950	Strychnine	1954
Patulin	1950	Quinine	1945
Lysergic acid	1954, 1956	Ergonovine	1954
Reserpine	1956		

* with revised structure

of his syntheses—reserpine and strychnine—will be discussed. Chlorophyll a was considered, but the complexity of the synthesis and gigantic size of the molecule make its inclusion cost too much space to include at the cost of other science in this and other chapters. Without question, there are many others that could be chosen instead. For example, it is fair to claim that Vitamin B$_{12}$ should be included here. I've chosen not to include this because it was completed after being awarded the Nobel Prize. That its size is on par with chlorophyll a only further supports my decision.

Before considering their total syntheses, some discussion of the chemical structures of reserpine and strychnine (Figure 5.1) is appropriate. Each of these targets presents its own set of synthetic challenges. Both contain multiple rings and ring fusions with several contiguous stereocenters. A modern reader may look at these structures and consider some of the targets pursued today as more challenging. Paclitaxel, brevotoxin b, and countless others at least match strychnine and reserpine. This, however true, is somewhat unfair and devoid of the context of Woodward's work. Well, understand that although Woodward had some analytical techniques we consider modern, much of the chemical structure elucidation that was done during the time was based on chemical analysis rather than instrumental analysis. Nevertheless, instrumental analysis certainly confirmed many of Woodward's chemical transformations. This alone increases the difficulty level of this sort of work for Woodward and all his contemporaries compared to modern synthetic efforts.

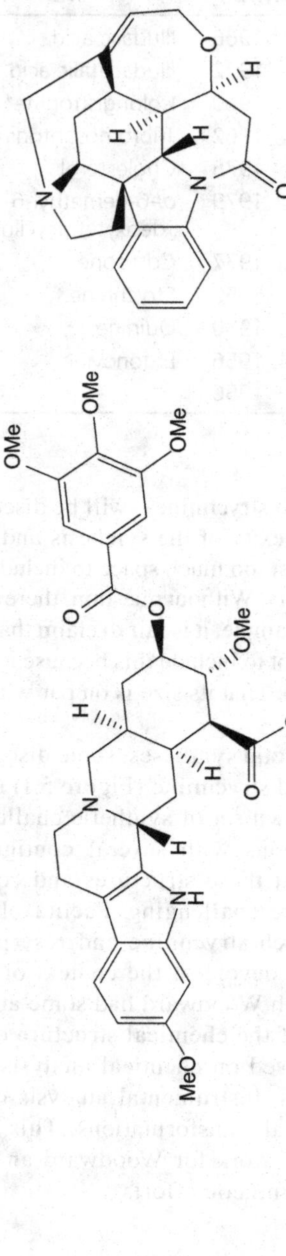

1

2

FIGURE 5.1: Chemical structures of reserpine (1) and strychnine (2).

CONTROVERSIES

There are at least two controversies (one of them ongoing) that have dogged Woodward after his death. His legendary smoking habit would certainly not be welcome today, but these controversies involve the science. One, regarding quinine, cast temporary doubt on one of his greatest accomplishments, and the other is one of integrity and a claim of misappropriating/not crediting a co-discoverer.

QUININE SYNTHESIS

Woodward and Doering reported *The Total Synthesis of Quinine* in 1944.[2] In this synthesis, Woodward and Doering actually did not prepare quinine, rather they synthesized a synthetic intermediate that had been shown to previously be an intermediate in the synthesis of quinine by Rabe. Today, we refer to this as a *formal* synthesis rather than a *total* synthesis. Not only does chemical technology change with the march of time, *but terminology does also* as well. Even rules for something as simple as chemical nomenclature change from time to time; I have had to change lecture slides and textbooks have had to change many a page to reflect new rules for naming compounds in the past few years. Unfortunately, years after Woodward's work, attempts to convert this intermediate to quinine failed. It's important to understand and recognize that these chemical steps were not those that were designed by Woodward and Doering, rather they were the work of Rabe.[3] A breakthrough came in 2008 when Williams and Smith[4] finally succeeded in the final and troublesome step of Rabe's method. Their breakthrough was as impressive as it was counterintuitive. They literally violated modern best practices and allowed an air-sensitive reagent to be exposed to the air prior to its use, essentially allowing it to spoil. They reasoned—and I dare say gambled—(correctly) that Rabe's 1930s lab would not have the modern controls we use today. By effectively replicating the laboratory conditions Rabe would have been in, they were able to complete the synthesis in a reasonable yield. It is, in truth, unfair to call this a controversy that is attributable to Woodward. After all, Woodward and his coworkers committed no foul at all by trusting that previously reported research would go as described. They did their part by accessing a key intermediate expediently. Moreover, even this once troublesome step is now resolved and verified, albeit unconventionally.

HOFFMANN AND COREY

A few years after Woodward's death, E.J. Corey, a Nobel laureate in his own right (1990), made a claim that can only be described as scandalous. Corey claimed (first in a letter to Hoffmann congratulating him on Hoffmann's 1981 Nobel Prize) that a conversation between Woodward and himself gave Woodward the inspiration or idea that led to the famed Woodward–Hoffmann rules, for which Hoffmann, together with Fukui was awarded a Nobel Prize for in 1981. Corey's claims were publicly made for the first time decades after Woodward's death, making it impossible for Woodward to defend himself against these claims. A new series is currently in the works exploring these claims. This controversy, however exciting it may be for everyone who loves a scandal, ends here for now. Decades later, there is no chance that Corey will be able to provide hard proof of his version of the events. There is no telling, however, what the aforementioned series will turn up.

SELECT SYNTHESES

Here, two of Woodward's syntheses will receive some spotlight. By no means am I meaning to insinuate that these are the "best two." I first restricted the focus to syntheses completed prior to Woodward's Nobel Prize. Reserpine and strychnine were chosen because of the enormous molecular complexity within both structurally and because both syntheses are just cool. Had Woodward's quinine synthesis been used to help the Allied Forces in World War II or been used in the global fight against malaria post-war, it would have without question been included in this volume. Perhaps future editions will. The synthesis of cholesterol, another (in) famous chemical, was also achieved by Woodward and perhaps deserves at least an honorable mention. I chose to exclude it from this chapter, however, for the sake of space. This was to allow for greater coverage of other science in other chapters; there is only so much space that can be devoted to large molecules and lengthy sequences for *one* chapter.

Woodward's synthesis of reserpine[5] (Figure 5.2) starts with a Diels–Alder cycloaddition. This is important for at least two reasons that it 1-sets three critical stereocenters, including a *cis* decalin ring fusion and also forms what is

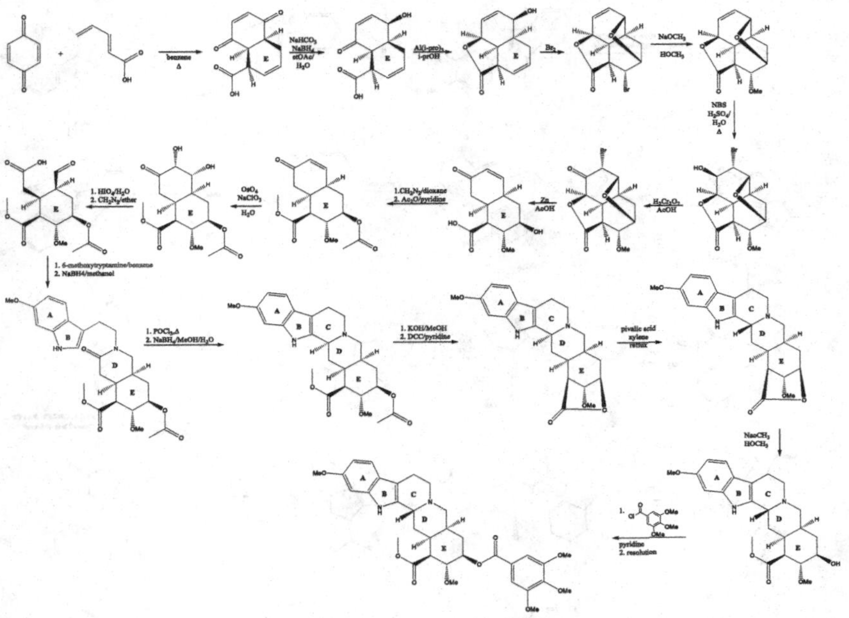

FIGURE 5.2: Woodward's synthesis of reserpine.

destined to become ring E. In the subsequent step, a selective reduction occurs, followed by a tandem reduction-lactamization sequence. The initial selective reduction Woodward reasoned was attributable to the proximal acid moiety. The synthesis then proceeds somewhat unremarkably, even if efficiently until the oxidative cleavage of the former quinone ring. This cleavage gives rise to an intermediate that allowed for the formation of the D-ring amide, proceeding through a Schiff base. This at once not only generates ring D, the use of a substituted tryptamine incorporates rings A & B with the appropriate decoration too. One of the greatest steps in this synthesis, however, is the one where the stereochemistry of the hydrogen atom at the ring C–ring D junction is inverted. This hydrogen atom is in the more thermodynamically stable position, conformationally. As a result, direct epimerization of it to the desired stereoisomer from this structure is not feasible. Woodward brilliantly solved this by tying the methyl ester (a protected acid) and an alcohol (masked as an acetate ester) to each other as a lactone. This brings about a conformational change for this hydrogen that allows it to be inverted since now, the inverted position is the one that is conformationally favored in this new structure.

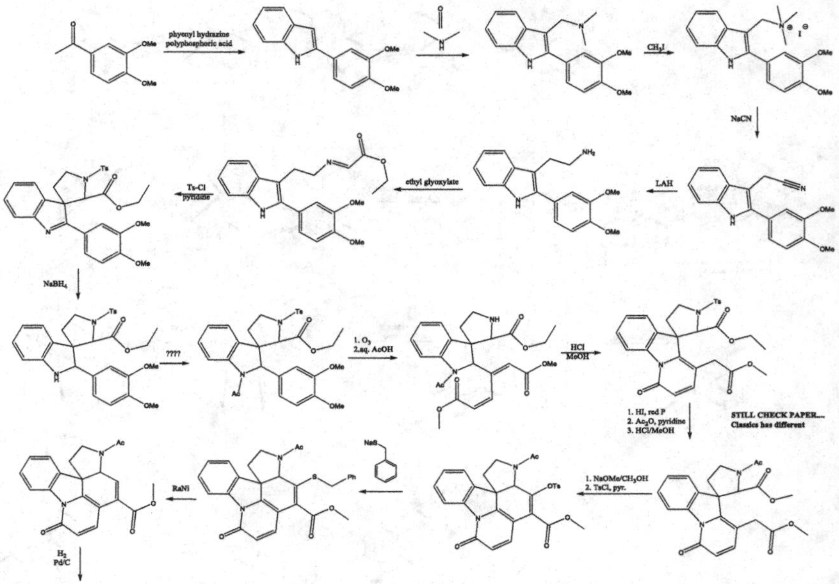

FIGURE 5.3:   Woodward's synthesis of strychnine.

Transformations such as this one is the bread and butter of what was considered the genius of R.B. Woodward.

The synthesis of strychnine[6] (Figure 5.3) starts with the formation of 2-veratoyl indole via a Fischer indole synthesis. The choice of the veratoyl group was very deliberate and key as it masks the alpha carbon. This decision allows for the eventual formation of ring V. It also easily withstood the reaction conditions it was exposed to. Woodward and his team foresaw that it could be selectively cleaved via ozonolysis on account of the electron-withdrawing nature of the two ester groups. This cleavage set the stage to generate ring III of strychnine. This cyclization takes place as part of an ester cleavage-lactamization-olefin isomerization, three transformation, one pot sequence. It is important to recognize that at this point, though rings still have some functional group modifications remaining, the skeleton for rings I, II, III, and V are now in place, a mere ten steps into the synthesis. Moreover, the presence of two esters in perfect relative position to make a requisite six-membered ring via a Dieckmann condensation is in place to generate ring IV. First, however, the stereocenter that the ester on ring V is attached to must be inverted. This ester is to be the electrophile in the aforementioned Dieckmann condensation, and in its current *trans* position, relative to the other ester, this condensation cannot proceed; they must be cis to each other to react. This required epimerization takes place in the same pot as the Dieckmann, but prior to it mechanistically. This epimerization was always a part of the plan. However, initial attempts to accomplish this went awry, destroying ring V. However, after some clever protecting group manipulation, a way forward was found that allowed for this epimerization while keeping ring V intact. Acetylation, followed by a base-induced rearrangement sets the stage for the creation of ring VI, the lactam. Lithium aluminum hydride then allows for a remarkable reduction where not only is the lactam of ring VI reduced to the amine, but a hydride is delivered to the quinone ring in the right stereochemical direction, presumably because of a coordinated complex and an ammonium cation. After this, an isomerization- and base-induced closure of ring VII furnishes the target strychnine.

SUMMARY

In closing, there is no really good way to in a short format demonstrate the full genius that merited Woodward's Nobel Prize. In fact, even in longer formats, it is difficult to capture since so much of what Woodward did is at least more common today than in his time. The reality is, however, that it is more common today because of the likes of Woodward and later Corey. These contributions are what made them both deserving of the Nobel Prize. Corey's award is covered in a later volume, but it can very rightly be considered to be part II of this chapter.

GENERAL SOURCES

https://www.nobelprize.org/uploads/2018/06/woodward-lecture.pdf, last checked 6/29/22
Classics in Total Synthesis 1ˢᵗ ed., E.J. Sorensen and K.C. Nicolaou. 1996, John Wiley
    and Sons, Inc.

CITED REFERENCES

1. https://cen.acs.org/articles/95/i15/Remembering-organic-chemistry-legend
   -Robert-Burns-Woodward.html
2. Woodward, R. B.; Doering, W. E. *Journal of the American Chemical Society,*
   1945, *67*, 860–874.
3. Rabe, P.; Kindler, K. *Berichte der Deutschen Chemischen Gesellschaft*, 1918,
   44, 2088–2091.
4. Smith, A. C.; Williams, R. M. *Ange ante Chemie. International Edition*, 2008,
   47, 1736–1740.
5. Woodward, R. B.; Bader, F. E.; Bickel, H.; Frey, A. J.; Kierstead, R. W.
   Tetrahedron, 1958, *2*, 1–57.
6. Woodward, R. B.; Cava, M. P.; Ollis, W. D.; Hunger, A.; Daeniker, H. U.;
   Schenker, K. *Journal of the American Chemical Society*, 1954, *76*, 4749–4751.

Index

A

acid chlorides, 61
alcohols, 44, 46, 56–57
aldehydes, 56, 57, 61
alkanes, 56
alkenes, 56, 58, 74
alkyne anion, 61
alkynes, 55, 56
amides, 57
amines, 56
awardees who declined the prize
    Butenandt, Adolf, 4
    Domagk, Gerhard, 4
    Kuhn, Richard, 4
    Le Duc Tho, 4, 16
    Pasternak, Boris, 4
    Sartre, Jean-Paul, 4

B

benzoic acid, 57
benzyl amine, 57
blue tungsten, 57

C

camphor, 52, 53
carbylamines, 57
cobalt, 55, 56
controversy
    Arafat, Yasser, 15, 16
    creation of the prize, 1
    Gajdusek, Carleton. D., 5, 16
    Haber, Fritz, 5, 16, 34
    Kissinger, Henry, 16
    President Barak Obama, 15
    Woodward, 94
cresols, 57

D

Diels–Alder reaction, 69, 70, 75, 78, 80, 82, 83, 86, 90
diene synthesis, 69
diketones, 57
diphenols, 57
diploma, 15

F

families
    Curie, 3, 5, 32, 33, 51
Felkin–Ahn model, 61
Fischer esterification, 46
Fischer indole synthesis, 44

G

Grignard reagent, 39, 55, 60, 62, 64, 66

H

hydrogenation, 52, 55, 56–59

I

incarcerated winners
    Aung San Suu Kyi, 5
    Liu Xiaobo, 5
    Ossietzky, Carl von, 5
iron, 55, 56
isocyanic esters, 57

K

ketones, 56–58, 61, 83
Kiliani–Fischer synthesis, 44

L

laureates, 3–5, 12, 14, 18, 22, 32, 67

N

Nazis, 15
nickel, 55, 56, 57
Nitriles, 56

O

organomagnesium, 60, 61
oximes, 57

P

Palladium, 58
pericyclic reactions, 69, 90
posthumous awards
    Hammarskjö, Dag, 3
    Karlfeldt, Erik Axel, 3
    Steinman, Ralph, 3

Q

quinine, 89, 93, 94
quinones, 57

R

reserpine, 78, 80, 91, 92, 95

S

snubs
    Bell Burnell, Joselyn, 20
    Carlson, Rachel, 20
    Franklin, Rosalind, 20
    Gandhi, Mahatma, 17
    Hawking, Stephen, 17
    Le Châtelier, Henry Louis, 17
    Lewis, Gilbert, 17
    Meitner, Liese, 20
    Mendeleev, Dimitri, 17
    Roosevelt, Eleanor, 20
    Sagan, Carl, 17
strychnine, 90, 92, 94, 96, 97

T

terpenes, 52–54
thorium oxide, 57

X

xylenols, 57

Printed in the United States
by Baker & Taylor Publisher Services